THE NATIONAL ACADEMIES

Advisers to the Nation on Science, Engineering, and Medicine

The **National Academy of Sciences** is a private, nonprofit, self-perpetuating society of distinguished scholars engaged in scientific and engineering research, dedicated to the furtherance of science and technology and to their use for the general welfare. Upon the authority of the charter granted to it by the Congress in 1863, the Academy has a mandate that requires it to advise the federal government on scientific and technical matters. Dr. Bruce M. Alberts is president of the National Academy of Sciences.

The **National Academy of Engineering** was established in 1964, under the charter of the National Academy of Sciences, as a parallel organization of outstanding engineers. It is autonomous in its administration and in the selection of its members, sharing with the National Academy of Sciences the responsibility for advising the federal government. The National Academy of Engineering also sponsors engineering programs aimed at meeting national needs, encourages education and research, and recognizes the superior achievements of engineers. Dr. Wm. A. Wulf is president of the National Academy of Engineering.

The **Institute of Medicine** was established in 1970 by the National Academy of Sciences to secure the services of eminent members of appropriate professions in the examination of policy matters pertaining to the health of the public. The Institute acts under the responsibility given to the National Academy of Sciences by its congressional charter to be an adviser to the federal government and, upon its own initiative, to identify issues of medical care, research, and education. Dr. Harvey V. Fineberg is president of the Institute of Medicine.

The **National Research Council** was organized by the National Academy of Sciences in 1916 to associate the broad community of science and technology with the Academy's purposes of furthering knowledge and advising the federal government. Functioning in accordance with general policies determined by the Academy, the Council has become the principal operating agency of both the National Academy of Sciences and the National Academy of Engineering in providing services to the government, the public, and the scientific and engineering communities. The Council is administered jointly by both Academies and the Institute of Medicine. Dr. Bruce M. Alberts and Dr. Wm. A. Wulf are chair and vice chair, respectively, of the National Research Council.

www.national-academies.org

U.S. SYMPOSIUM PARTICIPANTS

CHRIS G. WHIPPLE, *chair*, ENVIRON Corporation, Emeryville, California
MICHAEL H. BERGIN, Georgia Institute of Technology, Atlanta
JUDITH C. CHOW, Desert Research Institute, Reno, Nevada
LEONARD LEVIN, Electric Power Research Institute, Palo Alto, California
PAUL F. SCHWENGELS, Environmental Protection Agency, Washington, D.C.
DANIEL SPERLING, University of California, Davis
MICHAEL P. WALSH, consultant, Arlington, Virginia
HUA WANG, World Bank, Washington, D.C.

Staff

JACK J. FRITZ, Senior Program Officer, National Academy of Engineering
CAROL R. ARENBERG, Managing Editor, National Academy of Engineering
JOHN BORIGHT, Executive Director, Policy and Global Affairs, National Research Council
AIMEE CURTRIGHT, Research Associate, National Research Council

URBANIZATION, ENERGY, AND AIR POLLUTION IN CHINA

THE CHALLENGES AHEAD

PROCEEDINGS OF A SYMPOSIUM

DEVELOPMENT, SECURITY, AND COOPERATION
POLICY AND GLOBAL AFFAIRS
NATIONAL RESEARCH COUNCIL
OF THE NATIONAL ACADEMIES

NATIONAL ACADEMY OF ENGINEERING
OF THE NATIONAL ACADEMIES

CHINESE ACADEMY OF ENGINEERING
CHINESE ACADEMY OF SCIENCES

THE NATIONAL ACADEMIES PRESS
Washington, D.C.
www.nap.edu

THE NATIONAL ACADEMIES PRESS 500 Fifth Street, N.W. Washington, DC 20001

NOTICE: The project that is the subject of this report was approved by the Governing Board of the National Research Council, whose members are drawn from the councils of the National Academy of Sciences, the National Academy of Engineering, and the Institute of Medicine. The members of the committee responsible for the report were chosen for their special competences and with regard for appropriate balance.

This study was supported funding from the National Academies. Any opinions, findings, conclusions, or recommendations expressed in this publication are those of the author(s) and do not necessarily reflect the views of the organizations or agencies that provided support for the project.

International Standard Book Number 0-309-09323-6 (Book)
International Standard Book Number 0-309-54604-4(PDF)

Copies of this report are available from the National Academies Press, 500 Fifth Street, N.W., Lockbox 285, Washington, DC 20055; (800) 624-6242 or (202) 334-3313 (in the Washington metropolitan area); Internet, http://www.nap.edu

Cover Photo: Jack J. Fritz

Printed in the United States of America

CHINESE SYMPOSIUM PARTICIPANTS

ZHONGXIAN ZHAO, *chair,* Chinese Academy of Sciences, Beijing
RUIXIAN CAI, Institute of Engineering Thermophysics, Chinese Academy of Sciences, Beijing
WEITANG FAN (CAE), China Energy Research Society, Beijing
XIAOYAN TANG (CAE), Center of Environmental Science, Peking University, Beijing
FOSONG WANG, Chinese Academy of Sciences, Beijing
XUCHANG XU, Department of Thermal Engineering, Tsinghua University, Beijing
LUGUANG YAN, Institute of Electrical Engineering, Chinese Academy of Sciences, Beijing
JIANCHAO ZHENG, Electric Power Research Institute of China, Beijing

MEMBERS OF CHINESE SECRETARIAT

YINGSHI WANG, Secretary General, Professor, Institute of Engineering Thermophysics, Chinese Academy of Sciences
JIE CAO, Bureau of International Cooperation, Chinese Academy of Sciences
JINGHUA CAO, Office of American and Oceanian Affairs, Bureau of International Cooperation, Chinese Academy of Sciences
ZHE JIN, Division of Energy and Mining Engineering, Chinese Academy of Engineering
CHUNJIE LIU, General Office of Academic Divisions, Chinese Academy of Sciences
QI TIAN, International Cooperation Department, Chinese Academy of Engineering

Staff

RUI YANG, University of Science and Technology of China, Heifei

Acknowledgments

We wish to thank the National Research Council and the National Academy of Engineering of the National Academies, the Chinese Academy of Engineering, and the Chinese Academy of Sciences for their financial support of this project. We especially thank the members of the Chinese Secretariat for arranging this workshop and providing assistance to U.S. participants. The field visit to energy installations in and around Beijing showed the progress that has been made and the high level of effectiveness of steps taken to control air pollution around the capital city. Like previous exchanges between the Chinese and U.S. academies, this workshop showed that the economic future of both countries continues to be highly dependent on energy and the sound management of associated air pollution. Therefore, it is important that we work together to address the issues raised in these proceedings.

We would like to thank all of the participants and authors who worked on this project. The papers represent the views of the authors and do not represent official policy of the U.S. or Chinese academies. The authors were at liberty to present information both governments can use in developing policies to address air pollution. These papers represent a first step toward a major consensus study by the Chinese and U.S. academies on urban energy policy.

This volume has been reviewed in draft form by individuals chosen for their technical expertise, in accordance with procedures approved by the National Research Council's Report Review Committee. The purpose of this independent review is to provide candid and critical comments that will assist the institution in making its published report as sound as possible and to ensure that the report meets institutional standards for quality. The review comments and draft manuscript remain confidential to protect the integrity of the process.

We wish to thank the following individuals for reviewing the report: David Allen, University of Texas at Austin, and Lee Schipper, World Resources Institute. Although they provided constructive comments and suggestions, they were not asked to endorse the content of the individual papers. Responsibility for the final content of the papers rests with the individual authors.

Contents

GLOBAL IMPACTS

SAMPLING AND ANALYSIS

THE POWER SECTOR

INSTITUTIONAL ISSUES

PUBLIC HEALTH

Introduction

JACK J. FRITZ
National Academy of Engineering

Energy use and air pollution have been synonymous in China for decades, especially in urban areas. In rural areas, air pollution is also common because a significant amount of industry that is highly dependent on coal is located in the countryside. Fifteen or 20 years ago in China's northern cities, such as Shenyang, air pollution was characterized by decreased visibility caused by high levels of particulates and sulfur dioxide (SO_2). Although conditions have improved in modern cities, such as Beijing and Shanghai, China still has three of the ten most polluted cities in the world and hundreds of cities that are not in compliance with the World Health Organization (WHO) air quality guidelines.

China is undergoing urbanization and industrial development on an unprecedented scale. More than 120 cities have populations of more than one million, and by the end of the twenty-first century, 10 to 20 cities will have populations of more than 10 million. Rapid urbanization will challenge governments at all levels, not only to provide basic services to growing urban populations, but also to modernize, to continue to develop economically, and to address environmental concerns, particularly air pollution, that result from rapid economic growth.

In October 2003, a group of experts met in Beijing under the auspices of the Chinese Academy of Sciences, Chinese Academy of Engineering, and National Academy of Engineering (NAE)/National Research Council (NRC) of the National Academies to continue a dialogue and eventually chart a rational course of energy use in China. The importance of pollution abatement as part of an energy policy, a fairly recent idea in China, is already a major theme in national planning. In fact, in response to growing clamor for change by an increasingly prosperous and involved public, pollution reduction has been singled out as a priority in China's *Agenda 21* document.

Chinese planners now recognize that the choice of energy supply affects not only public health, but also land use, the environment, infrastructure, services, and economic growth. Thus, a secure, flexible, and varied energy-supply policy is critical to continued growth. Because China has an overabundance of coal and a scarcity of oil and gas, planners must continually balance the public good (i.e., public health and quality of life) against the easy availability of polluting coal and the high cost of importing oil and natural gas. Fundamentally, the Chinese policy community must address ambient air quality concerns by integrating energy supply and use for all economic sectors—industrial, power generation, residential, commercial, and transportation.

A good deal of progress has been made in China since the mid-1990s. The national averages for emissions of SO_2 and particulate matter (PM) have decreased, mostly as a result of stepped up enforcement of existing standards by national, provincial, and municipal governments. However, because of the increase in vehicle pollution and the continued prevalence of fine-particle pollution (less than 10 microns [PM_{10}], or even 2.5 microns in diameter [$PM_{2.5}$]), the government passed a second amendment in 2000 to the 1987 Law of Air Pollution Prevention and Control. The new legislation, which went into effect September 1, 2001, calls for the regulation of transportation, as well as residential and commercial energy use. When the new law is fully implemented over the next decade, it will greatly strengthen environmental laws and standards.

KEY TRENDS

One purpose of the October 2003 meeting was to identify trends that will influence future energy choices in China. These trends are discussed below.

The Presence of Fine Particulates (PM_{10} and $PM_{2.5}$)

In response to increased vehicle density and traffic congestion, China is implementing new control systems on combustion engines in cars, trucks, and small vehicles with two-cycle engines in hopes of reducing ground-level ozone and suspended particulates. To determine the effectiveness of these measures, China must first improve its monitoring of $PM_{2.5}$ and PM_{10}, as well as of gases, such as ozone. In a detailed report on source apportionment by Zhang et al., the authors note that their monitoring studies show the smallest particles in Beijing are predominantly from stationary sources (e.g., coal combustion and fugitive dust) rather than vehicles, despite the increase in vehicular traffic. This may be because vehicles in Beijing tend to be new and have fairly efficient combustion systems. This conclusion is affirmed by Xu et al. in their discussion of the power sector. The authors of both papers conclude that the monitoring and analysis of

PM and gases should be improved to ensure that policy makers have accurate data on the amount and sources of pollution. A related paper by Bergin adds that high aerosol loadings decrease visual range and attenuate solar radiation, which may result in decreased crop yields in nearby rural areas.

Chow and Watson discuss problems with sampling techniques and the unreliability of conclusions based on incomplete data. They also note the relationships between combustion-related particulates, fugitive dust, and precursors. For example, it has been widely assumed that the source of much of the PM in Xian is the desert west of the city. However, sampling reveals that PM is mostly from combustion sources and local dust, such as unpaved roads and empty tracts of land. Based on this information, policy makers can now develop a more effective, locally based strategy for controlling pollution in Xian.

Substituting Natural Gas for Coal

Natural gas is widely considered a viable replacement for coal, both for industry and home heating. But natural gas is expensive and not easily available in Chinese cities. In addition, further research will be necessary before vehicles that run on natural gas can be developed and before coal-fired heating boilers can be converted to natural gas.

Therefore, the Chinese government is committed both to the development of cleaner coal technology and to reducing the country's dependence on coal. Coal combustion continues to be the largest contributor to air pollution in China, with particulates and SO_2 causing the most significant problems. Although emissions of SO_2 and particulates have declined in some major cities since the mid-1990s as a result of improved controls and the increased use of low-sulfur coals in the power sector, emissions of both pollutants must clearly be reduced further. However, it is not obvious how this can be done economically. China needs a full analysis of SO_2 reduction under various control and policy strategies across various economic sectors. Xu et al. describe several control scenarios (with the required investments) for the power sector.

Xu et al. and Fritz both argue that coal will continue to be the dominant fuel for the next 50 years and that a variety of new technologies will be necessary to mitigate the negative environmental effects of coal consumption. These authors present a sober, realistic assessment of the difficulty of making a rapid transition to natural gas and renewable energy.

Based on the expectation that coal use will not only continue, but will even increase, Jin et al. make a strong case for the development and implementation of integrated gasification combined cycle (IGCC) with carbon dioxide (CO_2) sequestration technology for power generation from coal. Fan and Yu provide a detailed look at the composition of the energy supply in China and urge local authorities to increase energy efficiency, use advanced technologies, and diversify the fuel mix.

Vehicular Emissions of Nitrogen Oxides, Carbon Monoxide, Ozone, and Lead

NAE/NRC and the Chinese Academy of Engineering published a study in 2003 on the impact of the growing number of private automobiles in China. The study offered several recommendations for coping with the increase in pollution and congestion, but did not include a detailed plan. The report also documented some steps that had already been taken. For example, as of July 2000, leaded gasoline has been banned, starting in Beijing. However, emissions of nitrogen oxides (NO_x), carbon monoxide (CO), and ozone (O_3) from vehicles continue to increase. If, as expected, the number of vehicles increases dramatically, new regulations will be necessary just to maintain current air pollution levels and avoid the formation of smog.

Based on a consensus of the estimated number of vehicles in China in the coming years, Walsh offers recommendations for the development of clean, energy-efficient vehicles. This is the only way, he argues, China can address the inevitable increases in pollution from the addition of millions of vehicles, even if they are new, more efficient vehicles.

Sperling and Lin discuss air pollution from a mode of transportation unique to China, Chinese rural vehicles (CRVs), which outnumber conventional passenger vehicles by 3 to 1 and are extremely inefficient. The challenge for the Chinese government is to develop policies that mitigate the negative environmental impacts of these vehicles while promoting their economic benefits. Yan and Wen outline an electricity-based approach to the problem of urban transportation. They highlight recent technological advances in electric rail, electric vehicles, and high-speed magnetic-levitation (maglev) trains and describe their benefits.

Pollution in Small Industrial Cities

Perhaps the most challenging problem facing urban China is how smaller, coal-based, industrial cities can manage air pollution. Prosperous cities, such as Beijing and Shanghai, have enough resources to make radical changes in their energy mix. But most people live in mid-sized cities that do not have the resources to make radical changes. Tang describes some characteristics of air pollution in these cities brought about by rapid economic growth.

Smaller cities have experienced increases in SO_2, TSP, and NO_x levels as the result of their overwhelming reliance on coal for residential and industrial uses. Data on trends in levels of CO, O_3, and lead are not available for many small cities.

Reducing Sulfur-Dioxide Emissions and the Long-Term Implications of Acid Rain

Levels of acid deposition in China, like levels of SO_2 emissions, have decreased slightly since the mid-1990s. Nevertheless, as Kan et al. point out in their paper, acid rain remains a dangerous problem, particularly for human health. The major method of reducing SO_2 emissions has been to require that power plants use cleaner, low-sulfur coal and install flue-gas desulfurization technology. Although reductions are progressing reasonably well in the power sector, SO_2 emissions from state-owned industries are essentially unchanged.

Air-quality control at state-owned facilities remains a challenge. Many state-owned enterprises are not only outdated and uncompetitive, but are also financially stressed. Requiring stricter air-pollution compliance for these enterprises would force many of them into bankruptcy. Exacerbating the financial situation of these enterprises is the need to maintain a safety net for labor. Even though many of these industries are being restructured and the facilities sold off, serious questions about their future environmental impacts have yet to be addressed.

Guttikunda et al. describe a model that has been used to predict increases in SO_2 levels from the transport of emissions from new stationery sources in Asia. The model provides an inventory of large sources based on a grid size of 80 km by 80 km. Today, new investments in power-generating facilities, for example, must be run through the model to determine their impact and transport potential before they are approved. In neighboring Japan, this type of analysis is used to confirm local measurements of acid-rain deposition, a major concern.

Along similar lines, Levin describes the long-range transport of air pollutants and approaches for tracking them spatially and temporally across the Pacific Ocean. Mercury is used as a tracer substance, or proxy, for other pollutants, because it results from coal combustion and tends to be conservative.

Need for Improved Monitoring

Current and past techniques for measuring air pollution have often led to inaccurate results. It is unclear how, or if, these results have been used in setting policy and determining compliance, but China has no quality assurance programs in place to determine the accuracy of measurements or the condition of sampling instruments. Two papers, one by Chow and Watson and one by Zhang et al., address these problems. The authors of both papers conclude that an in-depth evaluation of equipment and methods should be undertaken, followed by an assessment of methods of data analysis and interpretation.

Positive Trends in Controlling Power-Sector Air Pollution

China has made significant progress in controlling pollution from the power sector, which now has many modern, state-of-the-art plants. Fritz argues that strict environmental oversight by the World Bank as a condition of loans to the power sector has brought pressure to bear on local environmental authorities to control emissions and procure state-of-the-art, pollution-control equipment. Questions remain, however, as to whether power-plant operators will continue to use the equipment. We know, for example, that in the cement industry, electrostatic precipitators are not used except when a monitoring cycle begins. Another problem is that policy makers lack a fundamental understanding of the fraction of ground-level air pollution that comes from the power sector and how much comes from small boilers. Most new power plants have stacks more than 200 meters high for wide dispersal, which makes it difficult to measure their contributions to ground-level pollution.

Emerging Regulatory Processes

The regulatory environment and compliance monitoring are the cornerstones of a system that can force polluters to reduce emissions. The paper by Wang and Wu addresses some of the institutional issues involved and introduces the reader to the Chinese regulatory framework. Although improvements in the current regulatory framework were not discussed in depth at the workshop, the group did reach a consensus that some of the techniques used in the United States might also be used in China, at least on a pilot scale. These include self-reporting, emissions trading, and tax-related incentives, coupled with tough enforcement.

Scale and Financial Impacts of Air Pollution on Public Health

The public health impacts of pollution are just beginning to be known. Some studies of hospital data and labor productivity have been undertaken, but at this point, the only conclusion that can be drawn is that the cost to the country and the Chinese people is enormous. Kan et al. take a preliminary look at methodologies for assessing health issues.

Whipple describes studies based on air-dispersion modeling of key sources in the United States to determine exposure to toxic substances emitted by power plants, such as mercury, arsenic, chromium, etc. To date, however, these studies have not provided a full accounting of health effects for these airborne species. Chinese authorities are monitoring U.S. progress in this area and are expected to adopt this type of monitoring in the coming years.

Emerging Technologies for Energy Generation and Distribution

As modernization in China continues, the private sector is looking for opportunities to become involved in power generation and distribution. Although China is still far behind other developing countries in opening the power sector to private operators, Feng and Wang describe a trend toward providing electricity, heating, and cooling power to new commercial ventures via cogeneration and distributed-energy systems. The trend is especially prevalent in private retail, industrial, and residential developments around Beijing and Shanghai. Fan and Yu identify areas for improvement, including the use of advanced power generation, combined heat and power generation (CHP), and non-coal fuel sources.

CONCLUSION

This collection of papers is intended to introduce the reader to the complicated problems of urban air pollution and energy choices in China. The positions of the authors do not represent official government policy or a consensus on best approaches. They do reflect recent thinking on the subject by individuals familiar with the issues. But we have only made a start, and much work remains to be done to bring additional Chinese and U.S. participants into the discussion.

The development of energy policy in China is important not only for China but also for the United States, Russia, India, and the European Union (EU). Each country or region has special concerns and limited "wiggle room" in terms of energy resources and economic resources, and each must make trade-offs between pollution abatement and costs. Clearly the United States and EU are better able on a per capita basis to shoulder these burdens than developing countries, which need cheap energy to move forward. Developed and developing countries have much to learn from each other. We hope that this gathering of experts on air pollution and energy will be the first in a series of interactions in the coming years.

BIBLIOGRAPHY

Chinese Academy of Engineering, National Academy of Engineering, National Research Council. 2003. Personal Cars and China. Washington, D.C.: The National Academies Press.

National Research Council, Chinese Academy of Sciences, Chinese Academy of Engineering. 2000. Cooperation in the Energy Futures of China and the United States. Washington, D.C.: National Academy Press.

World Bank. 2001. China: Air, Land, and Water. Washington, D.C.: World Bank.

Emerging Air Pollution Trends in China

Motor Vehicle Pollution and Fuel Consumption in China[1]

MICHAEL P. WALSH
Consultant, International Motor Vehicles

China has one of the fastest growing fleets of motor vehicles in the world. Since the late 1970s, the number of vehicles in China has increased about 10-fold. By the end of 2001, the total number of vehicles had reached about 18 million (excluding motorcycles), including 5 million cars (State Statistical Bureau China, 2002).[2] In 2000, China produced 2.07 million motor vehicles (a 43 percent increase from 1995), 605,000 of which were passenger cars (an 86 percent increase over 1995), and 11.53 million motorcycles, or 44 percent of the world's total production (an increase of 45 percent over 1995). By 2002, auto production had reached the 1 million mark, and indications are that the 2 million mark will be reached in the next year or two.

In early 2001, the Chinese government designated the automotive industry one of seven "pillar industries" of the economy. The government's Tenth Five-Year Plan proposes specific actions for restructuring and strengthening the automotive industry, which is now primarily engaged in manufacturing trucks and participating in joint ventures with foreign manufacturers for automobile assembly and the production of a Chinese family car at a price that will encourage mass ownership. Priority in the Tenth Five-Year Plan is given to investments in highways and oil and gas pipelines (State Economic and Trade Commission, 2001).

---

[1]This paper is based on an article published in December 2003. Motor vehicle pollution and fuel consumption in China: the long-term challenges. Energy for Sustainable Development 4: 28–39.

[2]The term "motor vehicles" in this report does not include two-wheeled vehicles or rural farm vehicles, unless otherwise indicated. It does include cars, trucks, buses, and commercial vehicles.

The rapidly growing automobile fleet has brought significant benefits to the Chinese people, including greater freedom of choice in housing location, employment, and leisure activities. But in the absence of government intervention, it also has continuing costs in urban areas: worsening air quality; an increase in the number of automobile accidents; and the toll on quality of life from congestion. Another cause for concern is the concomitant rise in energy consumption, which will mean China will become more dependent on imported petroleum.

FUTURE INCREASE IN VEHICLES

The three primary factors leading to increases in vehicle fleets in China, as in most countries, are population growth, urbanization, and economic improvement. According to the United Nations, the global population increased from approximately 2.5 billion people in 1950 to 6.3 billion today, and it is projected to increase to almost 9 billion, about 50 percent, by 2050. As Table 1 shows, this growth will not be evenly distributed but will be concentrated in Asia, Africa, and Latin America. China, already the most populous country in the world with 1.3 billion people, is expected to grow to 1.445 billion by 2025 before tapering off to 1.395 by 2050.

Urbanization will also continue in all regions of the world, with the greatest increases in southern Asia. This is significant because per capita vehicle populations are greater in urban areas than in rural areas because urban incomes are generally much higher than the national average. In China, for example, average incomes in Shanghai are three to five times higher than the national average. Thus, ownership of private cars is likely to be concentrated in Chinese cities. As Figure 1 shows, the number of cities in China with more than 200,000 people increased substantially during the 1990s.

TABLE 1 Worldwide Population Growth (in millions)

Region	1950	1998	2050
World	2,521	5,901	8,909
More developed regions	813	1,182	1,155
Less developed regions	1,709	4,719	7,754
Africa	221	749	1,766
Asia	1,402	3,585	5,268
Europe	547	729	628
Latin America and the Caribbean	167	504	809
North America	172	305	392
Oceania	13	30	46

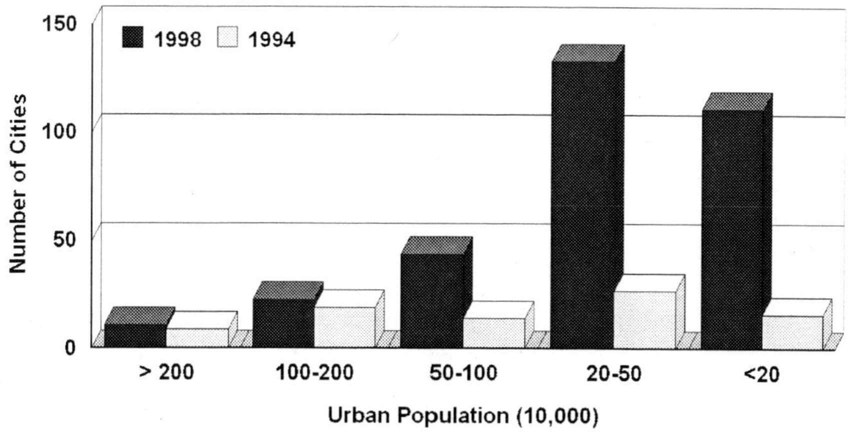

FIGURE 1 Urbanization trends in China. Source: He, 2001.

According to the Organisation for Economic Co-operation and Development (OECD), the increase in gross domestic product (GDP) in the next two decades will be highest in China, followed by east Asia, central and eastern Europe, and the former Soviet Union. Rising GDPs will be paralleled by increases in vehicle populations in these regions (personal communication, Peter Wiederkher, OECD). Since the 1960s, studies of the factors that influence the number of motor vehicles in countries and cities over time have consistently found that per capita income (as measured by GDP) is a major determinant of the size of the motor vehicle fleet (see Ingram and Liu, 1999, for a survey of these studies). On the national level, income alone typically explains more than 90 percent of the variation in motorization levels, and at the urban level more than 80 percent. Thus, with a growing population, rapid urbanization, and a rapidly growing economy, the vehicle population in China can be expected to grow steadily and substantially (Figure 2).

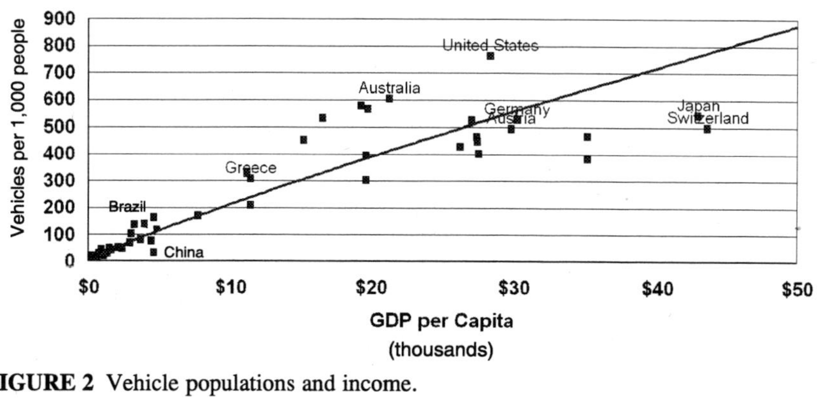

FIGURE 2 Vehicle populations and income.

PROJECTIONS OF CHINA'S MOTOR VEHICLE FLEET

The strong correlation between income and motorization provides a simple basis for projecting the size of the motor vehicle fleet in China. China's GDP, an indicator of personal income, grew at an average annual rate of 10.1 percent from 1980 to 1990 and 10.7 percent from 1990 through 1998. However, in 1999 and 2000 it grew at 7.1 and 8.0 percent, respectively (World Bank, 2001).

Study by the National Academy of Engineering

In 2003, the U.S. National Academy of Engineering and the Chinese National Academy of Sciences published a major study of the Chinese automobile industry (CAE et al., 2003). This report attempted to develop a first-order approximation of the likely vehicle population out to 2020 based on three different assumptions about the growth of China's GDP: a high growth rate of 10 percent; a medium growth rate of 8 percent; and a low growth rate of 6 percent. The projections are shown in Table 2. The medium-rate projection is close to

TABLE 2 National Vehicle Fleet Projections at Different GDP Growth Rates (millions of vehicles)

Year	Cars	Motor Vehicles
	10 percent GDP growth	
2005	7.9	26.4
2010	13.9	42.5
2015	24.4	68.4
2020	43.1	110.2
	8 percent GDP growth	
2005	7.2	24.5
2010	11.4	36.0
2015	18.0	52.9
2020	28.5	77.8
	6 percent GDP growth	
2005	6.6	22.7
2010	9.3	30.4
2015	13.2	40.7
2020	18.7	54.5

NOTE: Projections (in millions of vehicles) assume that total motor vehicle growth is the same as income growth and that automobile growth is 1.2 times income growth. GDP = gross domestic product. Source: CAE et al., 2003.

projections made earlier by the Chinese Academy of Engineering (State Economic and Trade Commission, 2001).

Study by Tsinghua University

In a second study, carried out primarily by Tsinghua University, the growth rate of GDP in China is projected to be 8 percent until 2010, 7 percent from 2010 to 2020, and 6 percent from 2020 to 2030 (He et al., 2004). The authors also attempt to account for expected changes in vehicle structure, based on government policies and development plans of the Chinese auto industry (Research Group of the Automotive Industry, 2001). The most important changes are listed below:

- explosion of the car population
- "dieselization," especially of trucks and buses
- shift from medium-sized to large trucks
- rapid growth in light-vehicle and minivehicle fleets

Figure 3 shows the estimated number of new vehicles by type for each year based on these changes; the vehicle population is forecast to increase by a factor of approximately 6 by 2030, with the most rapid increase in the car population. Figure 4 shows the relative ratios of cars to trucks and buses and the changing composition of the vehicle fleet; the proportion of cars is expected to reach

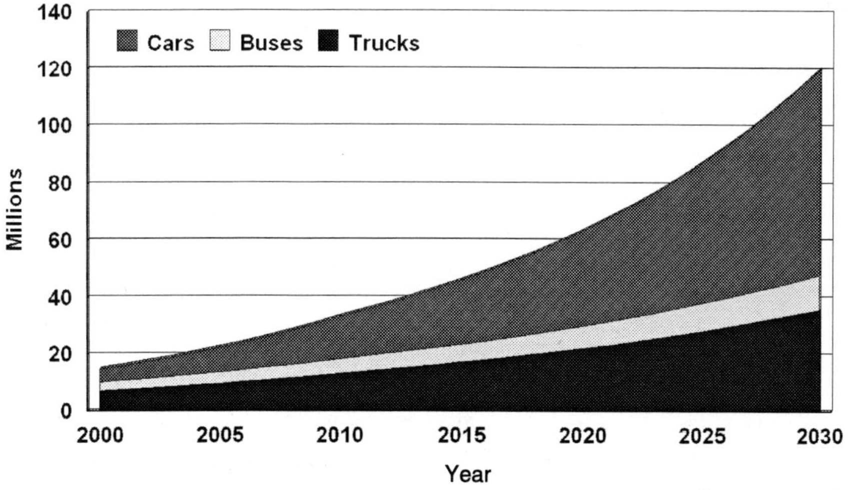

FIGURE 3 Forecast of total Chinese vehicle population (not including motorcycles). Source: Tsinghua University, 2002.

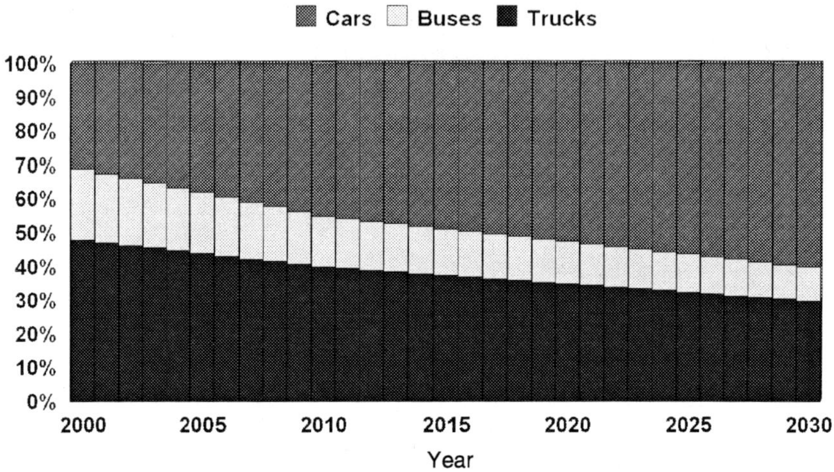

FIGURE 4 Forecast of Chinese vehicle population composition (not including motorcycles). Source: Tsinghua University, 2002.

60 percent by 2030. Figure 5 shows the trend toward higher diesel penetration, especially in the truck and bus fleets. Figure 6, a comparison of the Tsinghua and National Academy of Engineering forecasts, shows that they are remarkably similar.

ENVIRONMENT AND HEALTH

One of the obvious consequences of a rapidly growing vehicle fleet is its effect on the environment, particularly in cities. The air in most of China's large and medium-sized cities is already unacceptably polluted; the largest cities are ranked among the most polluted in the world. As Figure 7 shows, during the 1990s, NO_x air quality exceedances increased rapidly as the number of vehicles increased. Further increases will certainly exacerbate the situation unless major efforts are undertaken immediately to reduce emissions per vehicle.

Emissions

The combustion of gasoline or diesel fuel in vehicle engines produces a variety of potentially harmful emissions. The amounts and types of emissions depend on a variety of factors, including engine design, operating conditions, and fuel characteristics. Evaporative hydrocarbon emissions—from refueling, spills on heated engine parts, and so forth—can also be significant.

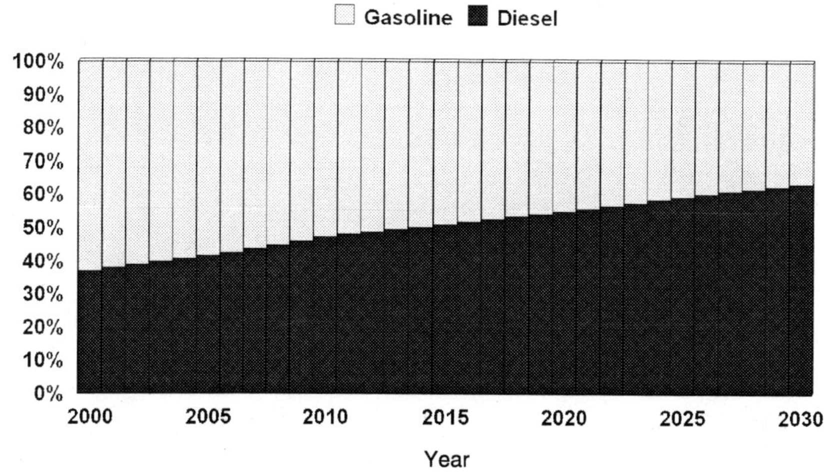

FIGURE 5 Forecasts of Chinese vehicle population fuel consumption (trucks and buses only). Source: Tsinghua University, 2002.

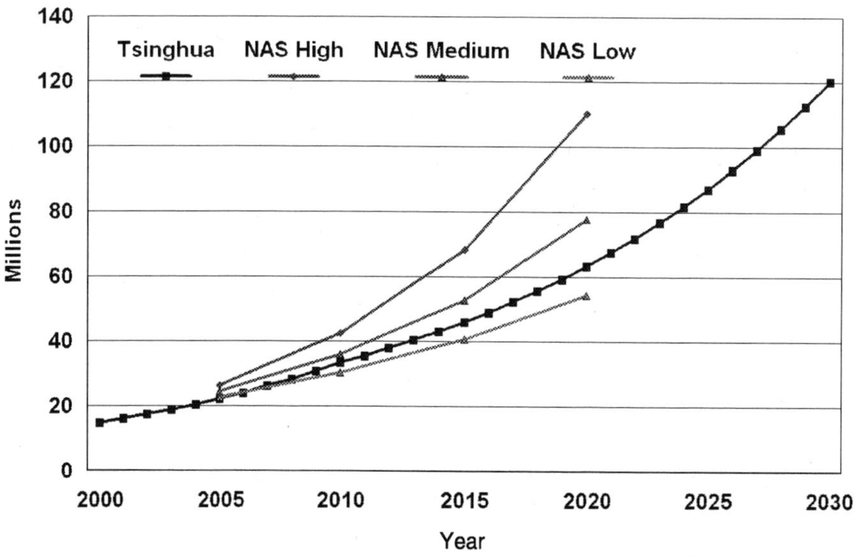

FIGURE 6 Comparison of the Tsinghua and National Academies forecasts of Chinese vehicle population (not including motorcycles).

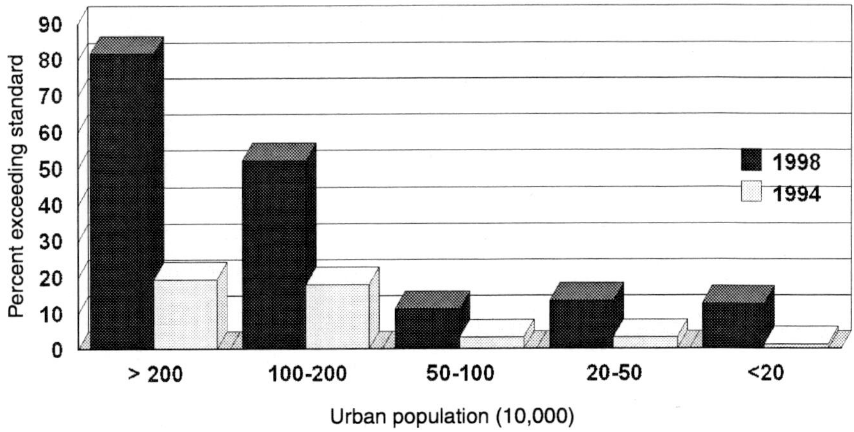

FIGURE 7 Rates of NO_x exceedences by city size, 1998 and 1994. Source: He, 2001.

The gaseous and particulate pollutants to which motor vehicles contribute include: carbon monoxide (CO); ozone (O_3)—through its atmospheric precursors, volatile organic compounds (VOCs) and nitrogen oxides (NO_x); fine particulate matter, PM_{10} and $PM_{2.5}$ (particles smaller than 10 and 2.5 microns [μm] in diameter, respectively); and nitrogen dioxide (NO_2). The toxic substances emitted from motor vehicles include aldehydes (acetaldehyde, formaldehyde, and others), benzene, 1, 3-butadiene, and a large number of substances known as polycyclic organic matter (including polycyclic aromatic hydrocarbons, or PAHs).

The relative contribution of motor vehicles to ambient levels of pollutants varies, depending on the pollutant and the location. In most cases, motor vehicles are a large and rapidly growing contributor to air pollution.

Health Effects

Research conducted over the past several decades has identified some of the effects of different pollutants on human health, including the respiratory, neurological, and cardiac systems and several types of cancer. Understanding the effects of a single pollutant can be difficult because usually pollution is a complex mixture of pollutants; and it is often difficult to disentangle the effects of a single pollutant from the effects of other pollutants that follow similar spatial and atmospheric patterns. At the same time, it is apparent that not all people are equally sensitive to the effects of pollutants; some subgroups (e.g., the elderly, asthmatics, children, people with heart disease) may be at greater risk from exposure to air pollution than healthy people (Kunzli et al., 2000). An air quality standard is now in place for NO_2, but before 2000, standards applied to both NO_2 and NO_x. Therefore, a good deal of the historical data are for NO_x.

Overall, the effects of pollutants on public health are of sufficient magnitude to be of great concern. For example, one recent European analysis estimated that approximately 6 percent, 40,000 deaths annually, in France, Austria, and Switzerland could be attributed to particulate air pollution alone, and about half of that could be attributed to exposure to vehicle emissions (Kunzli et al., 2000). A more recent study by the World Health Organization concluded that approximately 800,000 premature deaths occur each year as a result of exposure to urban air pollution, primarily particulate matter (WHO, 2002).

Air Quality

In spite of significant advances in industrial pollution control, air pollution in major Chinese cities remains a serious problem and, in some cases, may actually be getting worse. In addition, the pollution has changed from coal-based pollution to vehicle-based pollution. According to available data, the national air quality standards for NO_x are currently being exceeded across large areas of China, including, but not limited to, high-traffic areas (Tsinghua University, 2002). Before 1992, the annual average concentration of NO_x in Shanghai was less than 50 micrograms per cubic meter ($\mu g/m^3$), which complies with the Chinese Class II air quality standard. But since 1995, the concentration of NO_x has gradually increased, from 51 $\mu g/m^3$ in 1995 to 59 $\mu g/m^3$ in 1997. As illustrated in Figure 8, the change coincided with a rapid, steady increase in Shanghai's vehicle population (Shanghai Municipal Government, 1999).

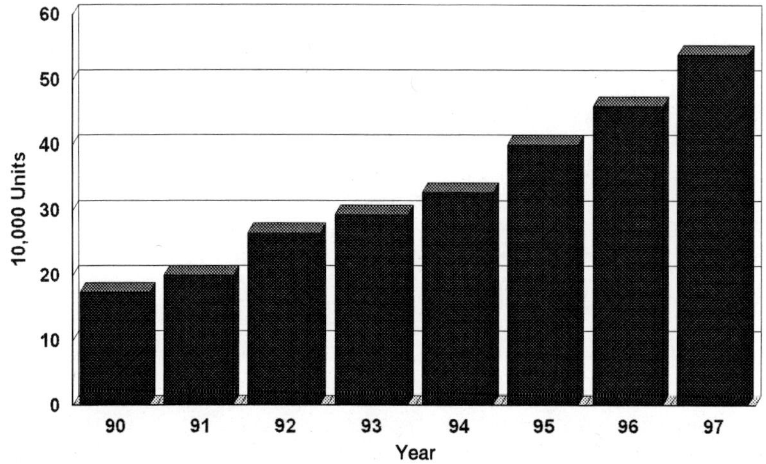

FIGURE 8 Vehicle population growth in Shanghai.

In Beijing, NO_x concentrations within the Second Ring Road, which encircles the city center, increased from 99 μg/m^3 in 1986 to 205 μg/m^3 in 1997, more than doubling in a decade. Moreover, concentrations of CO and NO_x on the trunk roads and interchanges exceed national environmental quality standards year round. In 1998, with the continued growth in vehicle population, concentrations of NO_x and CO in high-traffic areas exceeded the national standards throughout the year. Table 3 compares levels of NO_x and CO in Beijing's high-traffic areas (city center) and lower traffic areas in 1997 and 1998.

During the Eighth Five-Year Plan (1991–1995), O_3 concentration in Beijing's suburbs exceeded standards on average 53.8 days and 294 hours, respectively. In 1997, the average was 71 days and 434 hours, and the maximum hourly concentration reached 346 μg/m^3. In 1998, as Table 4 shows, the standard was exceeded on 101 days for a total of 504 hours, with the peak level rising to 384 μg/m^3. In 1999, O_3 levels exceeded the standard for 119 days and 777 hours.

The most serious air pollution problem in Chinese cities by far is particulate matter (PM), primarily from coal burning. A recent study, however, at two sites in Beijing found that vehicle exhaust now accounts for about 9 percent of $PM_{2.5}$, and re-entrained road dust accounts for approximately 14 to 15 percent. Samples were collected simultaneously at two sites in Beijing, one located at Chegongzhuang in a downtown area and one located on the campus of Tsinghua

TABLE 3 Concentrations of NO_x and CO in Urban and Suburban Beijing

	Year	2nd Ring Road (City Center)	3rd Ring Road	4th Ring Road	Outside 4th Ring Road (suburb)
NO_x (μg/m^3)	1997	205	190	177	112
	1998	220	219	197	124
CO (mg/m^3)	1997	6.8	6.1	3.3	
	1998	8.4	7.3	3.6	

Source: Tsinghua University, 2002.

TABLE 4 Changes in O_3 Pollution in Beijing

Year	Days Exceeding Standard	Hours Exceeding Standard	Max. Hourly Concentration of the Whole Year (μg/m^3)
1997	71	434	346
1998	101	504	384
1999	119	777	

Source: Tsinghua University, 2002.

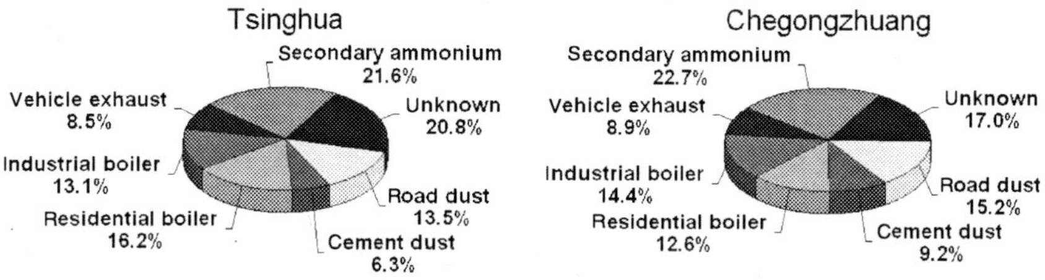

FIGURE 9 Average source contribution to $PM_{2.5}$ at two sites in Beijing. Source: He et al., 2002.

University in a residential area. The results are shown in Figure 9. It is worth noting that at the time of this study there were virtually no diesel cars in Beijing (they were banned until recently), and heavy diesel trucks were only allowed into the city at night (He et al., 2002).

STEPS TO ADDRESS POLLUTION

Improvements in Fuel Quality

Growing concerns in China about the environmental impacts of rising oil consumption have led to investments in new refining technologies and revisions in product specifications. One of the first policy targets was eliminating the 66 and 70 motor octane numbers (MONs) for gasoline, raising the new minimum to 90 research octane number (RON), and eliminating alkyl-lead additives for boosting octane through the addition of alkylate, reformate, and methyl tertiary-butyl ether (MTBE) and other oxygenates in gasoline blending. New unleaded specifications for 93 and 95 octane (RON) gasoline were added as well. Methyl cyclopentadienyl manganese tricarbonyl (MMT) is now used as an octane enhancer by about 50 percent of China's refineries.

Gasoline Specifications

Table 5 gives China's newest gasoline specification, GB 17930-1999 (China State Bureau of Quality and Technical Supervision, 1999). The total amount of olefins and aromatics in the gasoline pool are limited to 40 percent by volume maximum; the limit of olefin content is 35 percent by volume maximum.

Diesel Fuel

The original state specification for diesel fuel, GB 252-1994, was replaced by GB 252-2000 in January 2002 (the same standards as were introduced in the

TABLE 5 Specification for Unleaded Petrol for Motor Vehicles, July 2000 (GB17930-1999)

Item	Limit		
Research octane number (minimum)	90	93	95
Anti-knock index (minimum)	85	88	90
Lead (μg/liter maximum)		0.005	
Sulfur (ppm maximum)		1,000	
Benzene (% by volume maximum)		2.5	
Aromatics (% by volume maximum)		40	
Olefins (% by volume maximum)		35	

Source: China State Bureau of Quality and Technical Supervision, 1999.

Eropean Union (EU) in the early 1990s) (see Table 6). GB 252-2000 includes one grade with a maximum sulfur content of 0.2 percent by weight. The minimum cetane number limit is 45, with an exception for diesel fuels made from naphthenic or paraffin-naphthenic crude oils, which have a minimum cetane number limit of 40. These specifications imply that the cetane number of diesel fuels containing catalytic cracking components has a minimum limit of 45 rather than 40. In addition, China Petroleum and Chemical Corporation (SINOPEC) has issued a city diesel fuel specification, Q/SHR 008-2000. In this specification, the maximum sulfur content is 300 parts per million (ppm), and the minimum cetane number is 50 without exception.

On October 1, 2003, a new voluntary diesel fuel specification, Automobile Diesel Fuels GB/T 19147-2003, was introduced (Table 7).

VEHICLE-DIRECTED MEASURES

Once leaded gasoline was eliminated, China followed up in 2000 with the introduction of Euro I standards for new cars and trucks. Recently, China also decided to introduce Euro II standards in 2004. (Beijing introduced Euro II standards in January 2003, a year earlier than the rest of the country; Shanghai followed in April 2003.) The standard allows 50 ppm for all diesel and gasoline sold in the EU in 2005, and fuels with a maximum of 10 ppm must be widely available by that year. All fuel must comply with a maximum limit of 10 ppm no later than 2009. In addition, several Chinese cities are upgrading their vehicle inspection and maintenance programs to reduce emissions further.

In the city of Beijing, which has the highest per capita vehicle population and the most serious motor vehicle-related air pollution in China, a mandatory vehicle retirement policy was strictly enforced. Some 38,000 vehicles were forced off Beijing's roads by the end of 1998. Among these were 14,000 microbus taxies

TABLE 6 Diesel Fuel Specification GB 252-2000

Brand number	10	5	0	–10	–20	–35	–50
Solidifying point (^oC max.)	10	5	0	–10	–20	–35	–50
Cold-filtering plugging point (^oC max.)	12	8	4	– 5	–14	–29	–44
Flashpoint Pensky Martens Closed Tester (PM) (^oC min.)	55	55	55	55	55	45	45
Cetane number (min.)	45	45	45	45	45	45	45
Distillation temperature (^oC)							
50% vol. recovered at max.	300	300	300	300	300	300	300
90% vol. recovered at max.	355	355	355	355	355	355	355
95% vol. recovered at max.	365	365	365	365	365	365	365
Viscosity at 20^oC (mm^2/s)	3.0-8.0	3.0-8.0	3.0-8.0	3.0-8.0	2.5-8.0	1.8-7.0	1.8-7.0
Particulates (% by mass)	0	0	0	0	0	0	0
Copper corrosion rating (50^oC, 3hr max.)	1	1	1	1	1	1	1
Ash (% by mass. max.)	0.01	0.01	0.01	0.01	0.01	0.01	0.01
Carbon residue on 10% distillation residue (% by mass. max.)	0.3	0.3	0.3	0.3	0.3	0.3	0.3
Acidity (mg KOH/100 ml max.)	7	7	7	7	7	7	7
Water (% by vol. max.)	trace	trace	trace	trace	trace	trace	trace
Sulfur (% by vol. max.)	0.2	0.2	0.2	0.2	0.2	0.2	0.2
Color number (max.)	3.5	3.5	3.5	3.5	3.5	3.5	3.5
Oxidation stability (insoluble) (mg/100 ml max.)	2.5	2.5	2.5	2.5	2.5	2.5	2.5

KOH = potassium hydroxide; I = iodine. Source: China State Bureau of Quality and Technical Supervision, 2000.

TABLE 7 Automobile Diesel Fuels GB/T 19147-2003

Brand number	10	5	0	–10	–20	–35	–50
Solidifying point (°C max.)	10	5	0	–10	–20	–35	–50
CFPP (°C max.)	12	8	4	–5	–14	–29	–44
Particulate matter (°C min.)	55	55	55	55	50	45	45
Ignition property (one of the following properties must be met)							
Cetane number (min.)	49	49	49	49	46	45	45
Cetane index (min.)	46	46	46	46	46	43	43
Density at 20°C (kg/m^3)	820–860	820–860	820–860	820–860	820–860	800–840	820–860
Viscosity at 20°C (mm^2/s)	3.0–8.0	3.0–8.0	3.0–8.0	3.0–8.0	2.5 –8.0	1.8–7.0	1.8–7.0
Cetane number (min.)	45	45	45	45	45	45	45
Distillation temperature (°C)							
50% vol. recovered at max.	300	300	300	300	300	300	300
90% vol. recovered at max.	355	355	355	355	355	355	355
95% vol. recovered at max.	365	365	365	365	365	365	365
Particulates (% mass)	0	0	0	0	0	0	0
Copper corrosion rating (50°C 3 hr) max.)	1	1	1	1	1	1	1
Ash (% by mass. max.)	0.01	0.01	0.01	0.01	0.01	0.01	0.01
Carbon residue on 10% distillation residue (% by mass. max.)	0.3	0.3	0.3	0.3	0.3	0.3	0.3
Water (% by vol. max.)	trace	trace	trace	trace	trace	trace	trace
Sulfur (ppm max.)	500	500	500	500	500	500	500
Oxidation stability (insoluble) (mg/100 ml max.)	2.5	2.5	2.5	2.5	2.5	2.5	2.5
Lubricity (HFRR scar dia. @ 60) µm max.	460	460	460	460	460	460	460

Source: China State Bureau of Quality and Technical Supervision, 2003.

that caused heavy pollution. Approximately 4,000 vehicles were physically destroyed at Capital Steelworks.

More than 80,000 vehicles registered before 1995 were required to install vacuum time-delay valves to reduce NO_x emissions, and 120,000 newer vehicles were retrofitted with three-way catalysts. Approximately 21,000 taxies and buses were converted to dual-fuel vehicles (mostly liquid petroleum gas and gasoline); and about 1,500 buses were converted to operate on compressed natural gas (CNG). Some 360,000 vehicles coming into Beijing were inspected, and 109,000 were turned back.

FUEL QUALITY

Unless there are additional improvements in fuel quality, it will be difficult to tighten new vehicle standards further. Sulfur is of particular concern. The rapid increase in vehicles has been a primary force driving China's shift from a net exporter to a net importer of petroleum, raising concerns not only about China's energy security and balance of payments, but also about the increasing strains on China's refineries. Up to now, the Chinese network of refineries for producing indigenous heavy, sweet crude oil was able to meet most of the country's demand for refined petroleum products. Chinese refineries were built to process relatively low-sulfur domestic crude oil, and they have limited hydrodesulfurization capacity. Imported crude oil, however, has a much higher sulfur content than domestic crude. Moreover, because China has decided to follow the pollution control strategies of the EU, fuel quality will have to be upgraded, which will require further reductions in sulfur. Table 8 shows the fuel specification standards now in effect in the EU for fuels sold in 2005.

TABLE 8 European Union Fuel Specification Limits

Petrol/Gasoline	2000	2005	Diesel	2000	2005
RVP summer kPa (max.)[a]	60	—	Cetane number (min.)	51	—
Aromatics (% by vol. max.)	42	35	Density 15°C kg/m^3 (max.)	845	—
Benzene (% by vol. max.)	1	—	Distillation 95% by vol. (°C, max.)	360	—
Olefins (% by vol. max.)	18	—	Polyaromatics (% by vol., max.)	11	—
Oxygen (% by mass max.)	2.7	—	Sulfur (ppm max.)	350	50/10[b]
Sulfur (ppm)	150	50/10			

[a]RVP = Reid Vapor Pressure; kPa = kilopascals (1 atmosphere of pressure equals about 100 kPa).

[b]A maximum limit of 50 parts per million (ppm) applies for all diesel and gasoline sold in the EU in 2005, but fuels with a maximum limit of 10 ppm must be widely available by that year. All fuel must comply with a maximum limit of 10 ppm by 2009 at the latest. Source: OECD, 1998.

The quality of fuels is inextricably linked to regulations for vehicle emissions. Because lower sulfur levels in gasoline and diesel fuel are preconditions for the introduction of advanced vehicle technologies that are able to comply with Euro III and Euro IV standards and beyond, China will have to substantially upgrade its refineries. In addition, for China to get the full benefit of stringent vehicle standards, it will have to use very low-sulfur fuels.

Conclusions

Pollutants emitted by vehicles are a large and growing source of air pollution in China. They already account for a substantial fraction of the emissions contributing to excessively high ambient levels of air pollution. Even with the currently adopted emissions standards, Euro II by 2004 and a 10 percent improvement in vehicle fuel economy as called for in the Tenth Five-Year Plan, emissions of all pollutants will still increase because of rapid economic growth. The shift to diesel fuels, although helpful from the standpoint of improving fuel economy, will put even greater pressure on urban air quality because of high NO_x and PM emissions from diesel-fueled vehicles.

In view of the very rapid growth in the vehicle fleet forecast for the next three decades, China's environment could be subject to severe strains with significant public health consequences unless vehicle technology is substantially upgraded and fuel quality is improved. In addition, unless there are substantial improvements in vehicle technology, fuel consumption and greenhouse gas emissions will increase dramatically.

FUEL CONSUMPTION AND CARBON DIOXIDE EMISSIONS

At the end of the twentieth century, China was the third largest oil consumer after the United States and Japan (He et al., 2002). Oil consumption in China, which has increased by 4 percent per year for the past 20 years, reached 210 MMT in 2000 (State Statistical Bureau China, 2002). Issues associated with increasing oil demand have already emerged, particularly the insufficiency of the domestic oil supply to meet future requirements for economic development. Because of limits on domestic oil-production capacity, China has been a net oil-importing country since 1993; the amount imported in 2000 was 70 MMT (Report on China Energy Development, 2001).

The rapid development of transportation in China is largely responsible for the increasing oil demand. Motor vehicles in China consume about 85 percent of the country's gasoline output and 42 percent of the diesel output. In 1995, China's demand for oil was 3.0 million barrels per day (mbd), or 147 MMT per year. By 2000, the amount was 4.5 mbd (220 MMT). The amount projected for 2005 is 5.2 mbd (250 MMT) (Chen, 2001). By 2010, the projected amount is 270–310 MMT of crude oil per year, but the domestic supply is expected to reach just 165–200 MMT per year. The deficit of 105–110 MMT will have to be imported.

In the next three decades, the demand for oil will increase dramatically because of the rapidly increasing vehicle population. According to the Tsinghua study, unless vehicle fuel economy improves, the demand for oil by China's road transport sector will increase at an average rate of about 6 percent per year and will reach about 363 MMT in 2030, more than five times the demand in 2000. CO_2 emissions will rise to 1.15 billion tons.

Clearly, improvements in vehicle fuel efficiency will be necessary to offset the high growth rate. A significant effort to develop a program to improve fuel efficiency is under way with support from the U.S.-based Energy Foundation.

In addition, several recent studies have been done to evaluate the potential for alternative fuels to meet China's multiple policy goals: reducing greenhouse gas emissions; protecting human health; improving air quality; and achieving greater diversity in fuels and energy sources. Many promising fuels and technologies, such as hydrogen-powered fuel cells, have the potential to reduce the environmental impacts of transportation in the future. Other alternative fuels may also be important.

CONCLUSIONS

China's vehicle population has grown tremendously over the past two decades, and all indications are that it will continue to grow quickly for the foreseeable future. Although many individuals will benefit greatly from increased mobility, absent government intervention, air quality will worsen, the number of automobile accidents will increase, and congestion will take a toll on the quality of life. In addition, energy consumption will increase, and China will become more dependent on imported petroleum, with all of the economic and energy security consequences that will entail.

Although China has taken great strides in recent years to reduce vehicle emissions, much more will have to be done. Key strategies should include the adoption of very low-sulfur fuels, parity with Europe in new vehicle standards by 2010 at the latest, and an aggressive fuel economy improvement program. In the long term, advanced vehicle technologies and renewable fuels will become increasingly important.

BIBLIOGRAPHY AND REFERENCES

CAE, NAE, and NRC (Chinese Academy of Engineering, National Academy of Engineering, and National Research Council). 2003. Personal Cars and China. Washington, D.C.: National Academies Press. Also available in Chinese from Chinese Academy of Engineering.

Chen, M. 2001. Proceedings of 2001 Conference on Crude Oil Information. Beijing: SINOPEC Crude Oil Information Center (in Chinese).

China State Bureau of Quality and Technical Supervision. 1999. Unleaded Petrol Specification for Motor Vehicles, July 2000. GB 17930-1999. Beijing: China State Bureau of Quality and Technical Supervision.

China State Bureau of Quality and Technical Supervision. 2000. Light Diesel Fuel Specification GB 252-2000. Beijing: China State Bureau of Quality and Technical Supervision.

China State Bureau of Quality and Technical Supervision. 2003. Automobile Diesel Fuels B/T 19147-2003. Beijing: China State Bureau of Quality and Technical Supervision.

He, K. 2001. Presentation at the 94th Annual Meeting and Exhibition of the Air and Waste Management Association, Orlando, Florida, June 2001.

He, K., Q. Zhang, Y. Ma, F. Yang, S. Cadle, T. Chan, P. Mulawa, C. Chan, and X. Yao. 2002. Source Apportionment of $PM_{2.5}$ in Beijing. Available from Fuel Chemistry Preprints. Washington, D.C.: American Chemical Society.

He, K., H. Huo, Q. Zhang, D. He, F. An, M. Wang, and M.P. Walsh. 2004. Oil consumption and CO_2 emissions in China's road transport: current status, future trends, and policy implications. Energy Policy, in press.

Hui, G.K. 2001. Projections of Growth in Chinese Vehicles. Presentation by Prof. Guo Kong Hui, May 2001, Beijing, China.

Ingram, G.K., and Z. Liu. 1999. Determinants of Motorization and Road Provision. Pp. 325–356 in Essays in Transportation Economics and Policy, edited by J. Gomez-Ibanez, W.B. Tye, and C. Winston. Washington, D.C.: Brookings Institution Press.

Kunzli, N., R. Kaiser, S. Medina, M. Studnicka, O. Chanel, P. Filliger, M. Herry, F. Horak, V. Puybonnieux-Texier, P. Quenel, J. Schneider, R. Seethaler, J.C. Vergnaud, and H. Sommer. 2000. Public-health impact of outdoor and traffic-related air pollution: a European assessment. Lancet 356: 795–801.

OECD (Organization for Economic Cooperation and Development). 1998. Directive 98/70/EC of the European Parliament and of the Council of 13 October 1998, amending Council Directive 93/12/EEC. Brussels, Belgium: European Union.

Report on China Energy Development. 2001. Report on China Energy Development. Beijing: China Measurement Press.

Research Group of the Automotive Industry. 2001. Automotive industry development in the Tenth Five-Year Plan. Machine-Electricity New Production Gazette 1-2: 50–57.

Shanghai Municipal Government. 1999. Strategy for Sustainable Development of Urban Transportation and Environment for a Metropolis with Coordinating Development of Transportation and Environment toward the 21st Century. Shanghai: Shanghai Municipal Government.

State Economic and Trade Commission. 2001. China's 10th Five-Year Plan for the Development of the Automotive Industry (2001–2005). June 26, 2001. Beijing: State Economic and Trade Commission. Translated by China Business Update.

State Statistical Bureau China. 2002. Statistical Yearbook of China. Beijing: China Statistics Press.

Tsinghua University. 2002. SEPA-Tsinghua Annual Report of 2002. Beijing: Tsinghua University.

World Bank. 2001. World Development Indicators. Washington, D.C.: World Bank.

WHO (World Health Organization). 2002. The World Health Report 2002: Reducing Risks, Promoting Healthy Life. Geneva: World Health Organization.

Yang, J., W. Song, L. Wang, and J. Wang. 1997. Analysis for Supply and Demand of World Crude Resources and Trends in China's Petroleum Industry. Proceedings of 15th World Petroleum Congress, Beijing. In Chinese.

Clean Air and the Electrification of Urban Transportation

LUGUANG YAN
XUHUI WEN
Institute of Electrical Engineering
Chinese Academy of Sciences

Although cities have been around for at least 5,000 years, urban residents accounted for only 2 percent of the global population as recently as 1800. In the past 200 years, however, the trend toward urbanization has increased dramatically, and with rapid globalization, it has accelerated at an unprecedented rate. Modern modes of transportation and communication have woven cities worldwide into a closely connected network. Today, about 3 billion people, about one-half of the world's population, live in cities (Table 1). It is estimated that urbanization will accelerate even faster in the future. By 2030, urban residents are expected to number about 5 billion people, or 60 percent of the total world population (UN-Habitat, 2001).

TABLE 1 Urbanization of the World

Year	Number of Urban Residents	Percentage of Population
1900	160 million	10%
2000	2.5 billion	42%
2006	3.2 billion	50%
Year	**Metropolises >1 million**	**Megalopolises > 10 million**
1950	34	–
1995	214	–
2003	325	20

Source: UN-Habitat, 2001.

Developing countries are urbanizing much faster than developed countries, and the most economically underdeveloped countries are urbanizing fastest of all. By 2020, about one-half of the population of industrializing countries will live in cities and towns, and by 2030 the number of urban residents will increase from the current 1.9 billion to 3.9 billion (UN-Habitat, 2001). This will pose enormous challenges for those countries.

URBANIZATION IN CHINA

Although the Chinese have been constructing cities for more than 4,000 years, China had only 86 cities in 1949 when the People's Republic of China was established. At that time, the urbanization level in China was only 10.6 percent, compared with the world average of 29 percent and the average in developed European countries and the United States of more than 60 percent. The urbanization process in China can be divided roughly into four stages (Table 2).

China is now in a new stage of urban development. At present, China has 668 cities, including 32 megalopolises with more than 1 million residents and 43 metropolises with 0.5 to 1 million residents. In addition, China has 16,992 towns. The total number of residents in cities and towns is more than 349 million, about 30.4 percent of the population. It is estimated that the people who live in metropolises will constitute the majority of urban residents in the future. By 2010, China will have nearly 100 metropolises with more than 0.5 million residents each, a total of 150 million people. Included in this number are more than 40 megalopolises with more than 1 million residents each, a total of about 100 million people.

DEVELOPMENT TRENDS

The trends in urbanization in China in the twenty-first century are summarized below:

- The development of metropolises and megalopolises will continue.
- Regional city agglomerations will develop rapidly, driven by multiple central cities.
- The number of medium-sized cities will increase.
- Small towns will develop, and, as they do, urban and rural areas will become integrated.
- Tertiary metropolitan industries will develop into leading industries that prop up the national economy. Medium-sized and small cities, particularly small towns, will continue industrializing, focusing on traditional, labor-intensive manufacturing industries.
- Strategies for sustainable development will impose strict requirements for urban construction.

TABLE 2 Stages of Urbanization in China

Year	Stage of Urbanization	Situation	Number of Cities	Urban Populations and Level of Urbanization
1949 to 1957	Restoration	Until 1952, China reconstructed old city quarters, rectified urban social orders, and intensified urban production. By the end of 1957, China had set up 11 new cities and was promoting urbanization.	176 (1957)	99.5 million, 15.4 percent of the total population
1957 to 1965	Major fluctuations	As a result of policy mistakes, worsening Sino-Soviet relations, and severe natural disasters in three successive years, urbanization dropped. As of June 1963, China had lost 26 million urban residents.		Falls to 14 percent in 1964
1966 to 1978	Stagnation	During the 10 years of the Cultural Revolution, especially in 1969, the urbanization level dropped dramatically to 12.2 percent (the 1952 level).	193 (1978)	170 million, 17.9 percent of the total population
1979 to 2002	Steady development	Since 1978, China has gradually implemented urban reforms. Since 1984, economic reforms focusing on cities have accelerated development of cities and towns.	668 (1998)	349 million, 30.4 percent of the population

Between now and 2020, China is likely to encounter a grim situation because of its excessively large rural population. In periods of economic shortfall, increases in farm output lead to higher incomes for farmers, but when markets become glutted, increases in farm output result in uncertain incomes for farmers. Surplus markets in developed countries are generally saturated when per capita gross domestic product (GDP) reaches US$4,000. However, China's markets will be glutted, which will cause the prices of farm products to fall, when per capita GDP reaches only about US$800.

Therefore, China must vigorously boost the urbanization process and accelerate the development of city agglomerations around metropolises and megalopolises. In addition, China must promote the development of small and medium-sized cities and small towns. However, rather than just focusing on quantity-based expansion, the priority must be on moderate expansion, highlighting particular features, improved quality, and comprehensive development. Only in this way can China realize its objective of 60 percent urbanization, essentially reversing the proportions of urban and rural populations by 2020. In some cases, farmers' incomes will fall because of rising production costs.

POLLUTION FROM MOTOR VEHICLES

The Challenge of Urban Traffic

China is a developing country with a rapidly growing economy and rapid urbanization that is constantly expanding city limits and dramatically increasing urban populations. Urban residents are spending more and more time in transit and traveling longer and longer distances. Rapid urbanization naturally leads to a rapid increase in vehicle ownership. According to the National Bureau of Statistics of China (2001), the number of motor vehicles (excluding military vehicles) on the Chinese mainland increased tenfold, from 1.8 million in 1980 to 18 million in 2001. Figure 1 shows the increase in motor vehicles on the mainland in the past 22 years. It is estimated that China will have about 25 million vehicles in 2005, a 10 percent average annual increase. Of these, about 8.5 million will be cars, 34 percent of the total. If the current trend continues, urban areas of China will have 14 million cars in 2010, 10 million of them in metropolises.

According to statistics from the Beijing Municipal Traffic Control Bureau, as of August 4, 2003, Beijing had more than 2 million vehicles, including 1.3 million private vehicles. If the current trend continues, car ownership in Beijing will catch up with ownership in other world metropolises, such as Tokyo and New York, in a few years. However, the inhabitants of Tokyo do not suffer much from traffic jams, even though there are 4 million cars in the city. By contrast, traffic in Beijing is disastrous and sometimes poses grave problems for the city. For example, the speed of buses in Beijing has fallen from 16.7 km/h in 1994 to 9.2 km/h. At peak hours each day, nearly one-fifth of intersections and

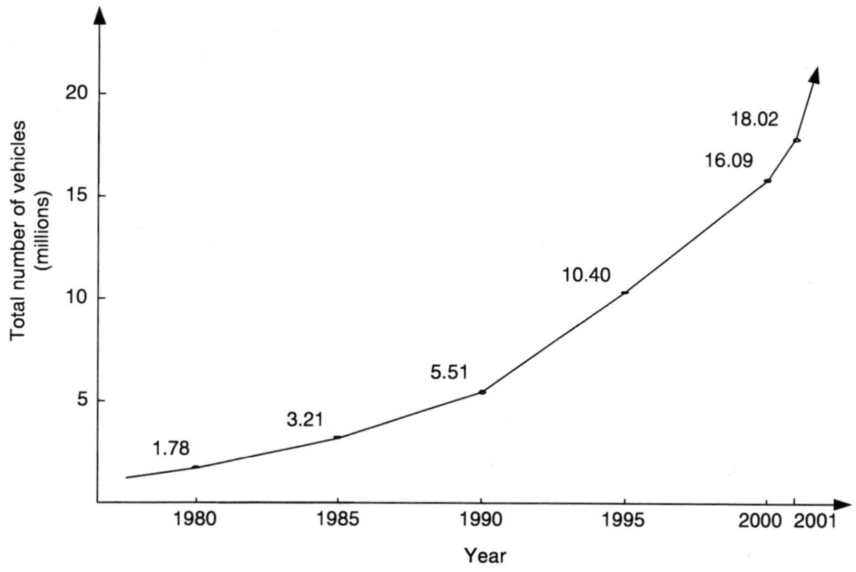

FIGURE 1 Increase in the number of motor vehicles on the Chinese mainland (excluding military vehicles), 1980–2001. Source: National Bureau of Statistics, 2001.

some sections of streets are choked with traffic, and cars proceed at a speed of 5 km/h. The same is true in some other cities in China.

Bicycles continue to serve as the basic mode of transportation for millions of Chinese people. China has 450 million bicycles.[1] Beijing alone has 9.28 million bicycles, 32 percent of which are used frequently. According to a recent survey on traffic, bicycles seriously interfere with the flow of vehicle traffic because they move slowly, occupy roads for a fairly long time, and prevent vehicles from speeding up quickly. Mixed traffic, particularly a mix of cars and bicycles, is a distinct feature of urban traffic conditions in China.

Impacts on Natural Resources

Rapid urbanization and the increase in the number of cars have led to significant shortages of oil and serious environmental pollution. China does have oil

[1]This is approximately four times the number of bicycles in the United States, roughly the same number per capita. The nature of bicycle use, particularly the frequency, is the distinguishing factor between the two countries.

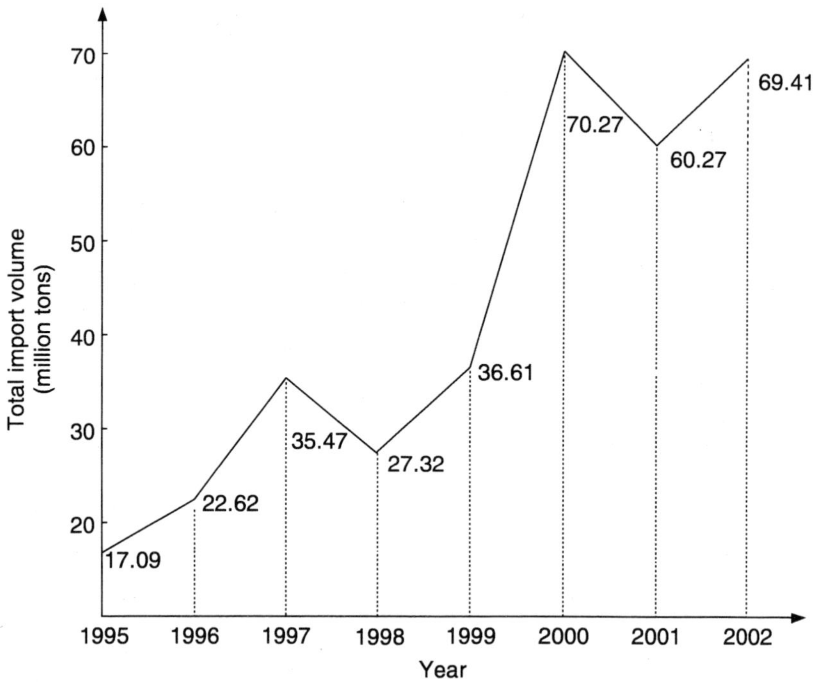

FIGURE 2 Oil imports in China, 1995–2002. Source: Customs General Administration, 2002.

resources. However, the ~20 billion barrels of oil account for only ~2 percent of current proven world reserves; China's oil production amounts to ~4.5 percent of total world output (EIA, 2004). Since 1993, China has gone from being a net exporter of oil to a net importer of oil. Figure 2 shows the change in China's oil-import status from 1995 to 2002.

Because domestic oil production is increasing slowly and the demand for oil is increasing rapidly, China's importation of oil can be expected to increase rapidly for the next 10 years. It is estimated that China will need 280 million tons of oil in 2005 and 360 million tons in 2010. For those years, China's domestic oil output will be 178 million tons and 189 million tons, respectively. China will, therefore, require net oil imports of 102 million tons and 171 million tons in 2005 and 2010, accounting for 36 percent and 48 percent of oil consumption, respectively. By 2020, China will consume 430 million tons of oil, requiring annual imports of 270 million tons of oil.

Motor vehicle fuels will make up an increasing proportion of China's oil requirements. At present, vehicle gasoline accounts for one-third of total gasoline consumption in the country. Although the government expects to develop alternative energy sources, particularly clean energy from coal gas and liquefied petroleum gas, growing oil demand will lead to many uncertainties in the development of the automobile industry in China.

Impact on the Environment

Scientific analysis shows that vehicle exhaust gas contains hundreds of compounds, including contaminants, such as solid suspended particles, carbon monoxide, hydrocarbons, oxides of nitrogen, lead, and sulfur-oxide compounds. According to some studies, vehicle tail gas accounts for 85 percent of air pollution on the Chinese mainland (Table 3).

Pollutants in vehicle exhaust gases can cause serious damage to people's health. Carbon monoxide combines with hemoglobin in human blood about 250 times faster than oxygen. Thus, people who inhale even a small volume of carbon monoxide are likely to suffer from severe anoxia. People exposed to small amounts feel dizzy and have headaches; people exposed to larger amounts can suffer permanent damage to brain cells. Oxides of nitrogen and hydrogen peroxide cause stimulus responses in susceptible people who suffer from ophthalmic diseases and laryngitis. Hydrocarbons from vehicle tailpipe emissions contain benzopyrene, a carcinogenic substance that forms particles that may be suspended in air for many days and, when inhaled, cannot be eliminated from the human body. If these substances accumulate in the lungs, they can cause cancer.

In addition to polluting the air, vehicles cause noise pollution. In 2001, traffic noise accounted for 20.1 percent of the overall environmental noise in the country. In districts with concentrated vehicle ownership, noise pollution is far

TABLE 3 Pollution from Vehicle Exhaust Gas

City	Pollution
Beijing (downtown areas)	Sulfur dioxide is 10–15 percent over the daily limit. Carbon monoxide and oxides of nitrogen are 60–70 percent over the daily limit. In extreme cases, air pollution is one to three times the national Grade II standard level.
Shanghai	Vehicle emissions of hydrocarbon account for more than 56 percent of total. Nitrogen fluoride is 20 percent above the limit.
Sichuan Province	Vehicles annually discharge 1.42 million tons of carbon monoxide and more than 600,000 tons of other pollutants. Vehicles contribute 80 percent of carbon monoxide and 90 percent of hydrocarbons.

higher. Vehicles also cause pollution through vibration and electromagnetic interference.

Some characteristics of vehicular pollution make it especially difficult to address:

- Because vehicles move around, the pollution source also moves, appearing mainly in downtown areas, factories, mines, government organizations, hospitals, and schools, where the population density is high. Vehicles can be considered "mobile chimneys" on the roads.
- Pollution from vehicles proliferates in belts rather than in points.
- Pollution density features periodical changes, keeping pace with people's activities; therefore, the most serious pollution occurs at peak traffic hours, when the concentration of people in urban areas is highest.
- Because it is much more difficult to shut down a street than a factory, it is difficult to stabilize the level of pollution.

ELECTRIFICATION OF TRANSPORTATION

The overall goals of clean energy and environmental protection are accepted by everyone, and the general trend is to work toward the development of vehicles that are pollution free, noiseless, and consume no petroleum. Many countries have implemented policies that advocate diversified energy sources to decrease reliance on petroleum and reduce oil consumption. Electric power can be generated from mineral resources (such as coal and natural gas), hydropower, wind power, tidal power, geothermal power, and nuclear power. As a result, electric power can save oil resources that are in comparatively short supply.

Electric-power-driven vehicles have the benefit of high efficiency and zero pollution at point of use. Thus, electrification is a rational choice for future development. At present, China has selected (1) electric vehicles, (2) electric rail traffic, (3) increased rail speed, and (4) the magnetic-levitation (maglev) trains as key development projects to address traffic, energy consumption, and pollution problems.

Electric Rail Systems

The most direct method of relieving urban traffic jams is to increase overall road capacity. However, metropolises generally do not have much space for new roads, so the traditional practice of new road construction to accommodate increasing numbers of vehicles is very ineffective. Urban electric rail systems take up little space and transport large numbers of passengers quickly. To date, the most widely used modes of electric rail traffic are metros (subways), suburban railways, and light-rail railways; these systems are the mainstays of passenger transport in many metropolises in the world.

China must clarify its policy toward the development of high-speed electric rail traffic, the most effective way to address traffic problems in big cities. In developing its urban electric rail traffic, China must give equal attention to above-ground and underground systems, connecting urban and suburban rail systems, and harmonizing rail traffic and other modes of transportation to create comprehensive, multifunctional, diversified city traffic systems based on rail traffic.

Electric Buses and Electric Bicycles

Various countries, particularly Western countries, are encouraging the development of electric cars to replace oil-based fuels with electric power. The Chinese government is also promoting the development of electric cars. In 1992, the development of electric cars was one of the key technological programs in the Eighth Five-Year Plan. But electric buses are ideal candidates for addressing urban traffic problems for the reasons outlined below:

- Considering current prices and the limited functionality of electric batteries, electric cars are not likely to satisfy the demands of private car owners in the immediate future.
- Electric buses can recharge batteries at times when it might be inconvenient for private owners to do so.
- Most large-scale bus lines can maintain buses by themselves, which eliminates the need for after-sale services.
- Bus-line purchasers generally figure vehicle costs on the basis of the lifetime of the bus, rather than simply first costs. This is an important element in comparing the economic benefits of electric vehicles and traditional fuel-based vehicles.
- Bus-line management can be encouraged to set socially and environmentally beneficial missions.
- Electric buses require a much smaller investment than electric rail systems.
- Electric buses can work efficiently in current urban environments.

As electric vehicle technologies and infrastructure systems mature, electric buses will inevitably replace traditional buses. Because electric buses cause no local air pollution and very little noise pollution and require low per-passenger energy consumption, they should become a major mode of urban transportation.

At present, electric vehicles are included in the "863" program of the Tenth Five-Year Plan, the main objectives of which are "industrialization of pure electric-driven vehicles, production in small batches of vehicles with mixed powers, and production of sample vehicles with fuel cells." Therefore, as electric bus technologies develop, electric cars will naturally eventually be used as well.

Electricity-driven bicycles will increase speed and reduce the physical burden on riders while maintaining the traditional advantages of bicycles—convenience and pollution-free transport. Electric bicycles will eventually become a favorite of urban residents for daily, short-distance transportation.

High-Speed, Magnetic-Levitation Trains

Modern metropolises generally have close connections with other large cities; they may also have one or more satellite cities. To provide transportation between cities and increase railway speed, China is also considering developing high-speed magnetic-levitation (maglev) trains. Maglev trains have several benefits: high speed; short acceleration time; strong capacity to climb slopes; and rapid braking. They are also very quiet and thus are ideal for intercity transportation. A maglev train system would greatly improve the efficiency of transportation among city agglomerations.

PROSPECTS FOR ELECTRIC VEHICLES IN CHINA

China has been working to develop electric vehicles for many years. From 1991 to 1995, the former State Science and Technology Commission and State Planning Commission included the development of electric vehicles in the Eighth Five-Year Plan. Tsinghua University and Tianjin Automotive Industry Company participated in government-sponsored research to develop a battery-powered, medium-sized bus and minicar. Studies have also been conducted on technologies for a sodium-sulfur battery, lead-acid battery, and permanent magnet DC motor drive. Some of the research results are being used in electric buses on the campus of Tsinghua University.

From 1996 to 2000, the State Science and Technology Commission carried out an Electric Vehicle Technology Industrial Project and a Fuel-Cell Technology Project. The results laid a solid foundation for the development of electric vehicles in China. For example, Dongfeng Automobile, together with the Institute of Electrical Engineering, Chinese Academy of Sciences, developed a concept electric sedan in 2000 (Figure 3). Lighter weight and with little wind resistance, the car had a digital, vector-controlled AC motor powered by a nickel-metal-hydride battery package. (Investigations were also made into lithium batteries.) A breakthrough was made in the development of a proton-exchange membrane (PEM) fuel-cell for electric vehicles, and a 30 kilowatt (kW) fuel-cell system was installed in a lightweight bus in 2001 (Figure 4). Table 4 shows comparative data for the sedan and bus. Meanwhile, an electric vehicle demonstration zone in Shantou City has been built for exploring the infrastructure and operation of electric vehicles.

Since 2001, when China entered the period of the Tenth Five-Year Plan, the Ministry of Science and Technology (formerly State Science and Technology

FIGURE 3 Battery-powered concept sedan, 2000.

FIGURE 4 Lightweight, fuel-cell-powered bus, 2001.

TABLE 4 Data for the Battery-Powered Concept Sedan and the Lightweight, Fuel-Cell-Powered Bus

	Maximum Speed	Maximum Climbing	Acceleration Time
Battery-powered conceptual sedan	114 km/h	20%	9.58 sec (0–50 km/h)
Lightweight, fuel-cell powered bus	60.6 km/h	18%	24 sec (0–40 km/h)

Commission) has included electric vehicle technology development as one of the "863" National High-Tech Special Projects. In five years, the country will have invested 880 million yuan in the development of electric vehicles and related technologies. Under the program, China will promote electric vehicle integration technologies, with a focus on batteries and fuel-cell engines, transmissions, and multi-energy source control for hybrid vehicles. Battery-powered electric vehicles will be developed for commercial operation in specific regions; hybrid electric vehicles will be targeted for small-batch production; and prototypes of fuel-cell vehicles will be produced.

With investments from the central government and local governments and the participation of a number of business enterprises, some important achievements were made in the first stage of the special project in 2002:

- Fuel-cell-powered engines were produced—30 kW for cars and 50 kW for buses. The system provides integrated, automated control of the various components (air supply, hydrogen supply, cooling, etc.).
- Progress was made on the development of a nickel-metal-hydride battery and a lithium battery (Table 5).
- AC motor drive systems provide power rated from 10 kW to 160 kW and maximum power density of 1.3 kW/kg, satisfying the requirements in terms of drive function and performance.
- Basic achievements have been made in multi-energy management technology for electric vehicles.
- Functional vehicles, including electric, hybrid-electric, and fuel-cell-powered cars and buses, have been successfully developed.

It is expected that by 2005, fuel-cell engines will be fairly well optimized to meet the power rating and system automation level requirements; significant efforts will still be necessary to improve power density. Prospects for using the high-power-density nickel-metal-hydride battery are good; the remaining

TABLE 5 Performance of Nickel-Metal-Hydride and Lithium Batteries[a]

Battery	Nickel-Metal-Hydride		Lithium	
	Domestic[b]	Global	Domestic[b]	Global
Energy density (Wh/kg)	65	~60–80	131	~100–150
Power density (W/kg)	680	~900–1,000	951	~1,500

[a]Nickel-metal-hydride batteries are better than lithium batteries in terms of life cycle, single-cell consistency, and security.

[b]The domestic data refer to a single cell.

problems are mostly related to costs. Lithium batteries will be used first for electric motorcycles, which require less power and energy, to explore the market. Research and development (R&D) on digital AC drive systems will be completed in this stage.

The next tasks for motor drive technology will focus on products and markets; research will continue on improving efficiency, lowering costs, and increasing power density. China is a major producer of rare earth materials, so considerable attention will be given to uses for permanent-magnet synchronous motors.

China will have the ability to produce small batches of hybrid electric vehicles and fuel-cell electric vehicle prototypes in 2005. However, electric vehicle industrialization and further commercialization will require government policy and legislative support. At present, we are not in a position to estimate the trends in electric vehicles in the overall Chinese auto industry.

HIGH-SPEED, MAGNETIC-LEVITATION TRAINS

The high-speed maglev train, developed in the second half of the twentieth century, is levitated by electromagnetic force. A long-stator, synchronous, linear motor drives the train at high speed. The train has no contact with the tracks and thus is not subject to the limitations of traditional wheel-track trains. The maglev train is currently the only land passenger transport with operating speeds of up to 500 km/h. It is expected that the maglev will be put into practical operation toward the middle of the twenty-first century. The maglev train system is best suited for long-distance, high-speed, intermetropolis transportation with large passenger volumes and will be competitive with air travel between large cities. It has the advantages of burning no fossil fuel, causing minimal pollution, and providing a safe, comfortable, economical, and reliable ride.

China began to develop the key technologies for a maglev train system in the late 1980s. The Railway Academy, Southwest Jiaotong University, National Defense Science and Technology University, and the Institute of Electrical Engineering of the Chinese Academy of Sciences conducted the work and successfully developed some test vehicles (Figures 5–7). In June 1998 at a joint conference of academicians of the Chinese Academy of Sciences and Chinese Academy of Engineering, Premier Zhu Rongji raised the question of using advanced maglev technologies in the construction of the Beijing-Shanghai high-speed railway. That discussion accelerated research into and discussion of a national development strategy for a high-speed maglev train system.

Taking into account the considerable gap between technical levels in China and the rest of the world, as well as the actual demands in China, a five-stage strategy was proposed: (1) study the need for the maglev train; (2) import technologies to construct a test operation line; (3) research and study the feasibility of long-distance lines and their application and establish an R&D team in China; (4) construct a long-distance line to realize a practical and industrialized

FIGURE 5 Six-ton, single-bogie truck maglev train developed primarily by Railway Academy.

FIGURE 6 Low-speed maglev train developed by Southwest Jiaotong University.

operation; and (5) gradually increase maglev trains in high-speed passenger transport networks. After considerable effort in recent years, China has made significant progress in developing the maglev train system. China plans to build a nearly 8,000 km, high-speed passenger transport network in the first half of the twenty-first century, and the maglev train system is the best candidate.

Work began on a test line in the winter of 1999; in the summer of 2000, China decided to cooperate with Germany in constructing a 30 km maglev train

FIGURE 7 Badaling tourism demonstration maglev train developed by National Defense Science and Technology University.

FIGURE 8 The Shanghai maglev train, Transrapid, in operation (430 km/h speed).

demonstration line from Pudong Airport to the urban area of Shanghai. The project officially started in March 2001 but was launched at a ceremony on December 31, 2002. After 22 months of construction, the first operating maglev train line was opened with a design speed of 430 km/h (Figure 8).

Construction of the Shanghai maglev line has accomplished several goals:

- It proved that a high-speed maglev train can run at a speed of more 400 km/h safely and reliably and that the technologies are mature enough for practical operation.

- The amazingly high speed of the system demonstrates that related technologies can be developed and mastered.
- It established the test base in China for R&D on high-speed maglev technologies.
- It brought together Chinese engineering, construction, and R&D teams.
- It enhanced Sino-German cooperation.

In addition, the construction of the Shanghai line demonstrated that the high-speed maglev train and the high-speed wheel-track train are both feasible candidates for the Beijing-Shanghai railway line. The next important task in the development of the high-speed maglev train is to persuade the government to use the maglev system for the Beijing-Shanghai high-speed line, for the following reasons:

- The high-speed maglev system is designed primarily for long-distance, intermetropolis, high-speed passenger transportation with heavy passenger volume. The Beijing-Shanghai line stretches 1,300 km. High-speed wheel-track transportation cannot compete with air travel in terms of travel time, but the high-speed maglev train system could connect the two cities in only three hours.
- Because of China's vast territory, large population, and short supply of oil resources, it is important that China develop new rail technologies. If the Beijing-Shanghai line adopts the maglev system, it will make maglev a leading technology in China's transportation system and will demonstrate China's success in developing new technologies.
- The adoption of the maglev system would stimulate industrialization of the total system and cultivate related high-tech industries.
- The maglev system produces little noise, consumes little energy, and demonstrates high performance in acceleration and deceleration.
- The maglev system could eventually be used for relatively shorter distances, including connections between multiple cities and intra-metropolis traffic lines. But "industrialization" is normally a precondition for promotion and popularization of a new technology.

BIBLIOGRAPHY AND REFERENCES

Customs General Administration. 2002. China's Customs Statistics (1995–2002). Beijing: Customs General Administration People's Republic of China.

EIA (Energy Information Administration). 2004. International Energy Annual 2002. Available online at: *http://www.eia.doe.gov/pub/international/iea2002/table81.xls* and *http://www.eia.doe.gov/pub/international/iealf/tableg2.xls* (accessed 7/22/04).

National Bureau of Statistics of China. 2001. China Statistical Yearbook 1981–2001. Available online at: *http://www.stats.gov.cn/english/*.

UN-Habitat (United Nations Center for Human Settlements [Habitat]). 2001. State of the World's Cities Report 2001. HS/619/01E. Nairobi, Kenya: UN-Habitat.

Wen, X., Z. Xu, G. Hu, Q. Zhang, G. Ma, and L. Zeng. 2001. Research on the Electric System of the PEMFC Test Minibus. In Proceedings of EVS-18 Symposium, Berlin, Germany, October 20–24, 2001.

Yan, L. 2000. China needs high-speed magnetically levitated train. Chinese Engineering Science 2(5): 8–13.

Yan, L., S. Xu, G. Su, Y. Dai, R. Zhang, and Y. Wu. 2002. Strategic progress of high-speed maglev and the development strategy in China (I). Advanced Technology of Electrical Engineering and Energy 21(4): 1–12.

Yan, L., S. Xu, G. Su, Y. Dai, R. Zhang, and Y. Wu. 2002. Strategic progress of high-speed maglev and the development strategy in China. Advanced Technology of Electrical Engineering and Energy 22(1): 1–8.

The Characteristics of Urban Air Pollution in China

XIAOYAN TANG
Center of Environmental Sciences
Peking University

China has experienced rapid economic growth (7 to 8 percent of gross domestic product [GDP] per year) since the mid-1980s. This rapid growth in such a short period of time has not only led to a remarkable increase in material wealth and a higher standard of living, but has also caused severe environmental pollution, particularly atmospheric pollution (He et al., 2002).

The problem was first observed in the 1970s with industrial emissions of sulfur dioxide (SO_2) and total suspended particulates (TSP). In the 1980s, acid rain was detected in major cities in the southern part of the country (Feng et al., 2002; Hao et al., 2001; Lei et al., 1997; Li and Gao, 2002; Qin and Huang, 2001; Tanner et al., 1997; Wang and Wang, 1995, 1996). This was caused mainly by SO_2 from coal combustion, which accounts for more than 70 percent of fuel consumption in China.

In the 1990s, the number of vehicles on roads increased very rapidly, especially in medium-sized and large cities. In Beijing, the number of vehicles increased by a factor of 4, from 0.5 million in 1990 to 2 million in 2002. In addition, the emission factor (the amount of pollution emitted by one car) in China is much higher than in developed countries, because China has much lower emissions standards for automobiles. Thus, the drastic rise in the number of vehicles and rapid development of industries in cities has led to worsening air quality (Xie et al., 2000, 2003; Zhang et al., 1999), particularly higher concentrations of nitrogen oxides (NO_x) (Wang et al., 2001) and particulates (Hu et al., 2002a,b; Kan and Chen, 2004; Song et al., 2002). High levels of ozone concentration were frequently observed in the summer and fall in several big cities (Ma, 2000; Tang et al., 1989, 1995; Wang et al., 2003; Zhang et al., 1998), and visibility in urban

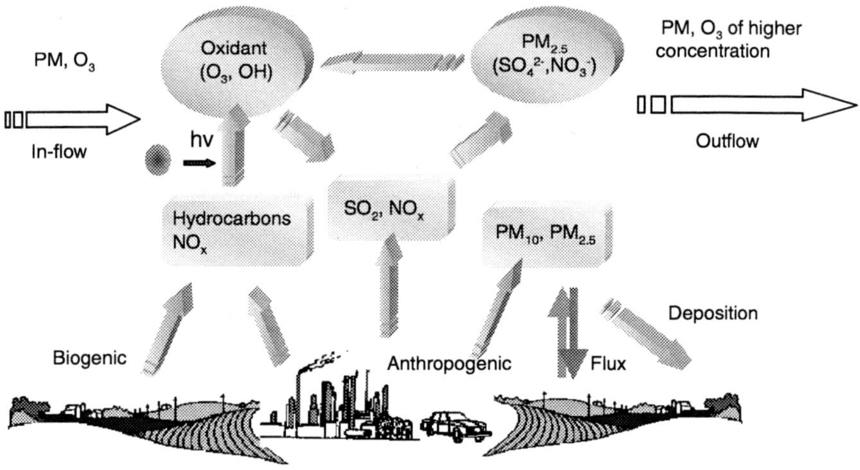

FIGURE 1 The chemical behavior of pollutants in the air.

areas continues to deteriorate (Song et al., 2003a). For the past three years, PM_{10} has been the predominant pollutant in most Chinese cities.

Economic growth and rapid urbanization in China have caused a tremendous increase in the consumption of energy and emissions of SO_2, NO_x (Hao et al., 2002), volatile organic compounds (VOCs) (Shao et al., 2000), TSP, and other pollutants. These primary pollutants not only disperse in the air and move to surrounding areas, but they also react photochemically to generate secondary pollutants. Photochemical smog (ground-level ozone) is produced from the reaction of NO_x and VOCs with ultraviolet radiation. Gas-phase SO_2, NO_x and VOCs can be transformed into fine particles ($PM_{2.5}$), which have large surface areas and can catalyze further reactions on the particle surface. Thus, reactions between pollutants are cyclical, leading to buildups of secondary pollutants in the air and causing severe air quality problems (Figure 1).

This phenomenon has been observed not only in several megacities, such as Beijing, Guangzhou, and Shanghai, but also in many medium-sized and large cities. This kind of air pollution has three distinguishing characteristics: (1) a high concentration of fine particles that adversely affect visibility; (2) a high capacity for atmospheric oxidation; and (3) regional environmental effects.

HIGH CONCENTRATIONS OF FINE PARTICLES AND ADVERSE EFFECTS ON VISIBILITY

The visibility problem has affected Beijing for several years (Song et al., 2003a). Citizens often complain that they rarely see blue skies or white clouds in

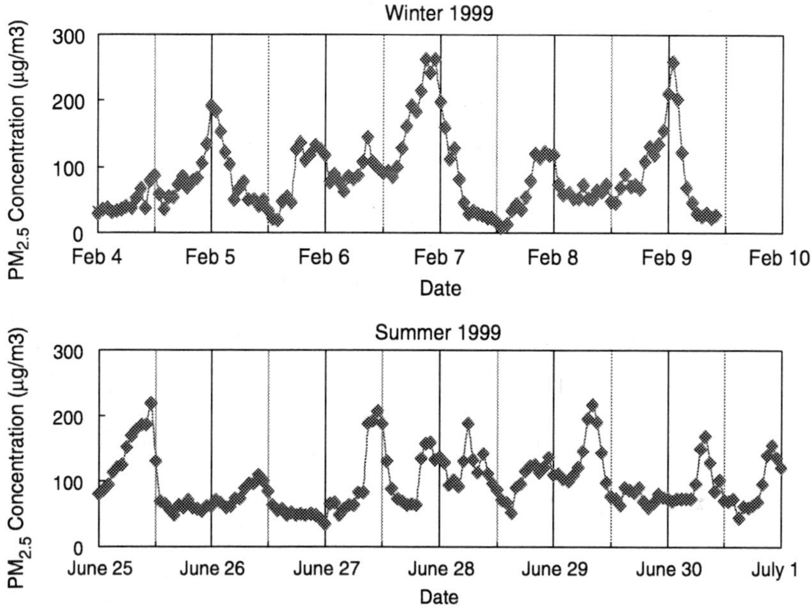

FIGURE 2 The diurnal profile of $PM_{2.5}$ mass concentration in winter (above) and summer (below) in 1999 in Beijing.

the Beijing area. A study of the relationship between visibility and concentrations of $PM_{2.5}$ in 1999–2000 showed a direct correlation in every season (Song et al., 2003b). Results also showed that the concentration of $PM_{2.5}$ in the summer and winter (a daily average of 60 to 80 μg/m^3) was higher than the national air quality standard in the United States (65 μg/m^3). In winter, the $PM_{2.5}$ concentration was highest at midnight (Figure 2), probably because heavy vehicles were permitted to pass through the metropolitan area only after 8:00 p.m.

HIGH CAPACITY FOR ATMOSPHERIC OXIDATION

Since the 1990s, high concentrations of surface ozone have been observed in Beijing from May to October. The diurnal profiles of ambient ozone concentration observed at the Peking University monitoring site for several years are shown in Figure 3 (Ma and Zhang, 2000; Zhang et al., 1998). The peak ozone concentration has increased every year, no doubt as a result of economic and population growth. In other megacities, such as Guangzhou and Shanghai, high ozone concentrations have also been observed in residential areas.

In June and July 2000, intensive measurements were taken of fine particles

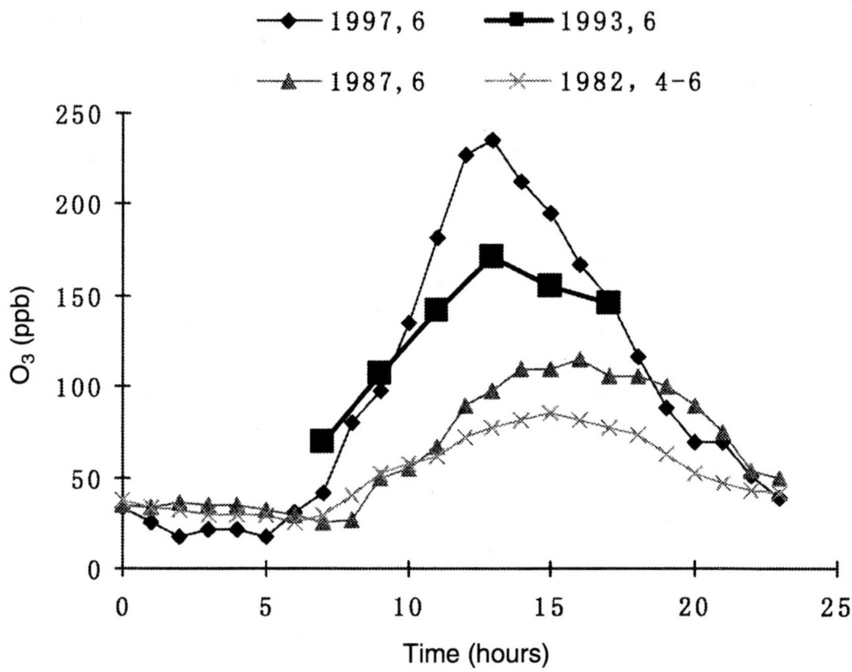

FIGURE 3 The diurnal profile of ozone concentration measured in different years at the Peking University site. Source: Zhang, 1998.

and ozone levels (and precursor levels) at six sites in the Beijing area. A high ozone episode occurred between June 21 to July 3, and the daily ozone peak (more than 200 ppb) occurred at about noon on most days.

Ozone as a secondary product is produced simultaneously with a series of oxidants, such as atmospheric free radicals (e.g., OH, HO_2, RO, RO_2, etc.). High concentrations of ozone and free radicals in the atmosphere lead to a high potential for the oxidation of primary pollutants (SO_2 and NO_x) to secondary pollutants (sulfate [SO_4^{-2}] and nitrate [NO_3^-]). The increased concentration of secondary pollutants is probably the major reason for the increasingly serious visibility problem.

Short-term and long-term exposure to ozone-rich air can cause eye irritation, plant damage, respiratory problems, and the deterioration of rubber and paint. The World Health Organization has suggested that one hour of exposure should not exceed 75 to 100 parts per billion (ppb). Most cities in China cannot meet that standard. Therefore, reducing the number and severity of ozone episodes and lowering the levels of atmospheric ozone have become major health concerns of city governments.

REGIONAL EFFECTS OF AIR POLLUTION

Because of geographic conditions, rapid economic growth has taken place mainly in the eastern part of the country, particularly in coastal areas. The Yangtze River delta region, the Pearl River delta region, and the Beijing-Tianjin-Bohai Bay region are typical of economically developed zones in China. In just 20 years, urbanization has increased dramatically in those regions, and the distances between cities have steadily decreased. As the density of cities has increased, the regional impact of air pollution has become increasingly noticeable.

In recent years, field measurements of air pollutants with meteorological observations and model simulations of air quality on a regional scale have been conducted in these three regions. The results showed pollutants, particularly secondary pollutants such as fine particles and ozone, were distributed regionally under certain meteorological conditions (Municipal Area Project Report, 2002; Pearl River Delta Region Project Report, 2002).

Figure 4 shows the regional effects of ozone in the Pearl River delta region measured in an intensive study (PGSAQPRD, 2002). A very strong negative relationship was found between concentrations of ozone and NO_x from upwind to downwind sites. At the site in the downtown area of Guangzhou, the ozone

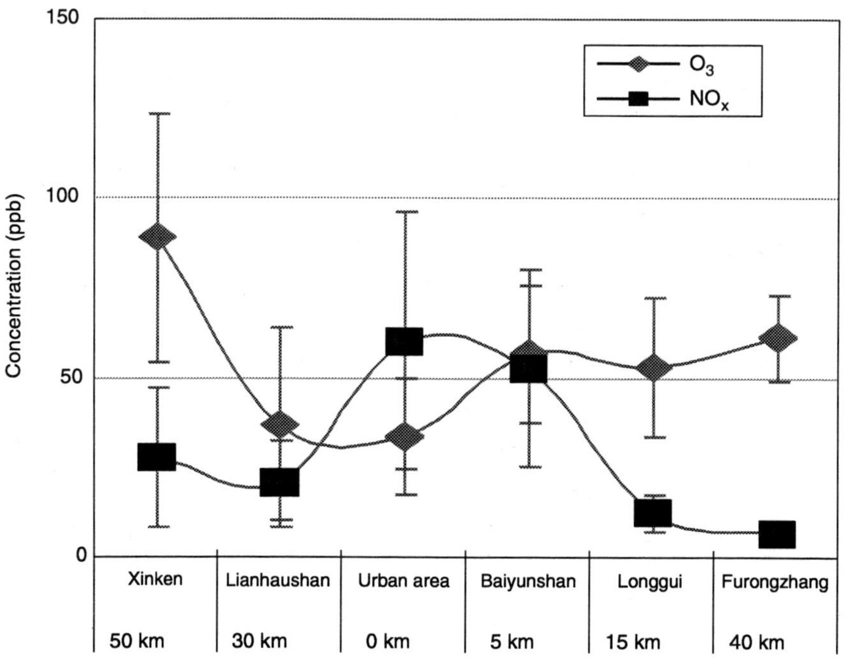

FIGURE 4 The upwind-to-downwind distribution of ozone and NO_x showing daytime average concentrations in Guangzhou and contiguous areas. Source: PGSAQPRD, 2002.

concentration was very low because of the titration of NO_x with ozone. At the downwind remote site (50 km), the ozone concentration was higher because of pollution transport by wind. The same phenomenon was found in the summer in the same region. In Beijing, similar research has also shown regional effects (PGAPC, 2002).

Figure 5 shows the regions included in the three-dimensional numerical simulation of air quality in Beijing and the surrounding area. The model shows that ozone episodes vary and the contributions to ozone formation come from different sources depending on meteorological conditions. The simulation results for June 27, 2000, are shown in Table 1. The daily maximum concentration of 200 ppb appeared at 1 to 2 p.m. in the Beijing downtown area; high ozone covered a large region, including part of Tianjin and surroundings. The model calculation indicates that the main contributor to the ozone episode at that time was precursors from Tianjin city and residential areas (Municipal Area Project Report, 2002).

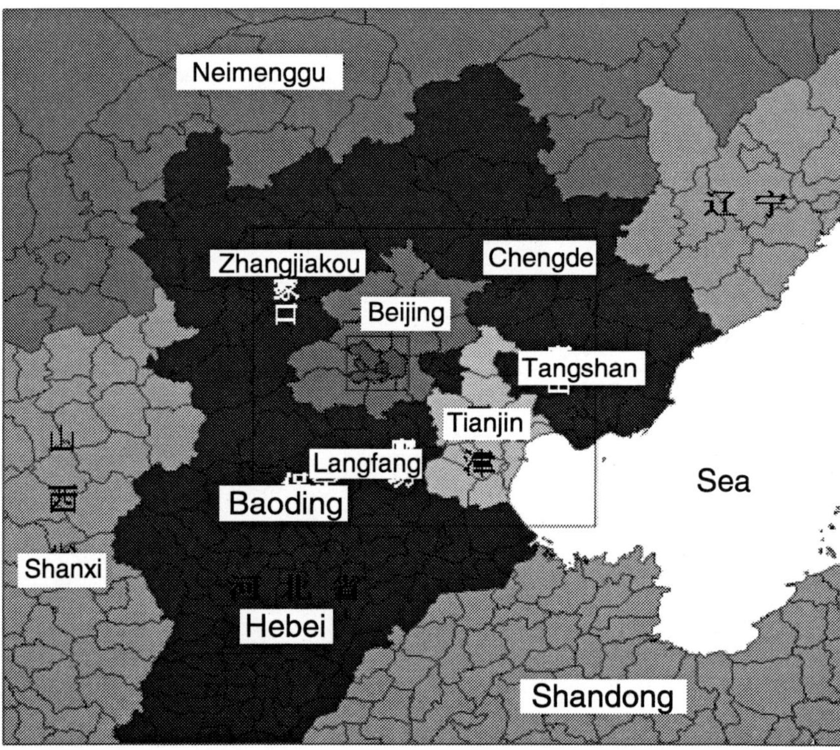

FIGURE 5 The simulation area, including Beijing municipality and surrounding areas, in the three-dimensional air quality model in three scales of nesting grids. Source: PGAPC, 2002.

TABLE 1 Contributions of Precursors (NO_x and VOCs) from Various Locations in the Formation of Ozone in Downtown Beijing (1–2 p.m., June 27, 2000)

Source Region	NO_x Contribution	VOC Contribution	Total Contribution
Beijing downtown	9.9%	14.7%	24.6%
Southeast part of Beijing	5.6%	3.5%	9.1%
Tianjin and vicinity	9.3%	23.8%	33.1%
Southern part of Hebei Province	3.3%	7.9%	11.2%

Source: Municipal Area Project Report, 2002.

CONCLUSION

Rapid economic growth combined with urbanization has caused severe air pollution problems in China, especially in urban areas. These problems are proving very difficult to solve because of the complex nature of the pollution. Studies have been focused on identifying the formation mechanisms, sources, and impacts on health and the ecosystem. Much attention has also been focused on control measures and policy issues. China expects that the air pollution problem will be reduced in the next 10 to 15 years.

REFERENCES

Feng, Z.W., H. Miao, F.Z. Zhang, and Y.Z. Huang. 2002. Effects of acid deposition on terrestrial ecosystems and their rehabilitation strategies in China. Journal of Environmental Sciences-China 14(2): 227–233.

Hao, J.M., H.Z. Tian, and Y.Q. Lu. 2002. Emission inventories of NO_x from commercial energy consumption in China, 1995–1998. Environmental Science and Technology 36(4): 552–560.

Hao, J.M., S.X. Wang, B.J. Liu, and K.B. He. 2001. Plotting of acid rain and sulfur dioxide pollution control zones and integrated control planning in China. Water Air and Soil Pollution 130(1-4): 259–264, Part 2.

He, K.B., H. Huo, and Q. Zhang. 2002. Urban air pollution in China: current status, characteristics, and progress. Annual Review of Energy and the Environment 27: 397–431.

Hu, M., L.Y. He, Y.H. Zhang, M. Wang, Y.P. Kim, and K.C. Moon. 2002a. Seasonal variation of ionic species in fine particles at Qingdao, China. Atmospheric Environment 36(38): 5853–5859.

Hu, M., F.M. Zhou, K.S. Shao, Y.H. Zhang, X.Y. Tang, and J. Slanina. 2002b. Diurnal variations of aerosol chemical compositions and related gaseous pollutants in Beijing and Guangzhou. Journal of Environmental Science and Health, Part A-Toxic/Hazardous Substances and Environmental Engineering 37(4): 479–488.

Kan, H.D., and B.H. Chen. 2004. Particulate air pollution in urban areas of Shanghai, China: health-based economic assessment. Science of the Total Environment 322: 71–79.

Lei, H.C., P.A. Tanner, M.Y. Huang, Z.L. Shen, and Y.X. Wu. 1997. The acidification process under the cloud in southwest China: observation results and simulation. Atmospheric Environment 31(6): 851–861.

Li, W., and J.X. Gao. 2002. Acid deposition and integrated zoning control in China. Environmental Management 30(2): 169–182.

Ma, Y.L., and Y.H. Zhang. 2000. The study on pollution of atmospheric photochemical oxidants in Beijing. Research of Environmental Sciences 13(1): 14–18.

PGAPC (Project Group on Air Pollution Control). 2002. Municipal Area Project Report: Strategic Study on Air Pollution Control in Beijing, 06/02. Paper by Project Group on Air Pollution Control. Unpublished.

PGSAQPRD (Project Group on Study on Air Quality of the Pearl River Delta). 2002. Pearl River Delta Region Project Report: Study on Air Quality of Pearl River Delta Region, 02/02. Paper by Project Group on Study on Air Quality of the Pearl River Delta. Unpublished.

Qin, G.Y., and M.Y. Huang. 2001. A study on rain acidification processes in ten cities of China. Water Air and Soil Pollution 130(1-4): 163–174.

Shao, M., M.P. Zhao, Y.H. Zhang, L.X. Peng, and J.L. Li. 2000. Biogenic VOCs emissions and its impact on ozone formation in major cities of China. Journal of Environmental Science and Health, Part A-Toxic/Hazardous Substances and Environmental Engineering 35(10): 1941–1950.

Song, Y., X.Y. Tang, Y.H. Zhang, M. Hu, C. Fang, L.M. Zeng, and W. Wang. 2002. Effects on fine particles by the continued high temperature weather in Beijing. Environmental Science 23(4): 33–36.

Song, Y., X.Y. Tang, Y.H. Zhang, M. Hu, C. Fang, and L.M. Zeng. 2003a. The study of the status and degradation of visibility in Beijing. Research of Environmental Sciences 16(2): 10–12.

Song, Y., X.Y. Tang, C. Fang, Y.H. Zhang, M. Hu, L.M. Zeng, C.C. Li, and J.T. Mao. 2003b. Relationship between the visibility degradation and particle pollution in Beijing. Acta Scientiae Circumstantiae 23(4): 468–471.

Tang, X.Y, J.L. Li, and Z.X. Dong. 1989. Photochemical pollution in Lanzhou, China: a case study. Journal of Environmental Sciences (China) 1: 31–38.

Tang, X.Y., J.L. Li, and D.H. Chen. 1995. Summertime photochemical pollution in Beijing. Pure and Applied Chemistry 67(8-9): 1465–1468.

Tanner, P.A., H.C. Lei, M.Y. Huang, and Z.L. Shen. 1997. Acid rain and below-cloud scavenging in south-western China. Journal of Atmospheric Chemistry 27(1): 71–78.

Wang, H.X., X.Y. Tang, M.L. Wang, P. Yan, T. Wang, K.S. Shao, L.M. Zeng, H.F. Du, and L.M. Chen. 2003. Characteristics of observed trace gaseous pollutants in the Yangtze delta. Science in China Series, D-Earth Sciences 46(4): 397–404.

Wang, S.X., M. Shao, and Y.H. Zhang. 2001. Influence of fuel quality on vehicular NO_x emissions. Journal of Environmental Sciences-China 13(3): 265–271.

Wang, W.X., and T. Wang. 1995. On the origin and the trend of acid precipitation in China. Water Air and Soil Pollution 85(4): 2295–2300.

Wang, W.X., and T. Wang. 1996. On acid rain formation in China. Atmospheric Environment 30(23): 4091–4093.

Xie, S.D., Y.H. Zhang, and X.Y. Tang. 2000. Current situation and trend of motor vehicle exhaust pollution in urban areas of China. Research of Environmental Sciences 13(4): 22–25, 38.

Xie, S.D., Y.H. Zhang, Q. Li, and X.Y. Tang. 2003. Spatial distribution of traffic-related pollutant concentrations in street canyons. Atmospheric Environment 37(23): 3213–3224.

Zhang, Y.H., K.S. Shao, and X.Y. Tang. 1998. The study of urban photochemical smog pollution in China. Acta Scientiarum Naturalium Universitatis Pek Inensis 34(2-3): 392–400.

Zhang, Y.H., S.D. Xie, L.M. Zeng, H.X. Wang, K.H. Yu, C.J. Zhu, N.M. Puan, and B.G. Wang. 1999. Traffic emission and its impact on air quality in Guangzhou area. Journal of Environmental Sciences 11(3): 355–360.

Rational Options for Clean Energy in Chinese Cities

WEITANG FAN
China Energy Research Society

ZHUFENG YU
Clean Coal Engineering and Research Center

Although renewable energy technologies are making rapid progress in the world, 20 to 30 years hence, coal, oil, and natural gas will still be the primary sources of energy in China. The proven recoverable reserves of fossil fuels in the world are shown in Table 1. China ranks third in coal reserves, eleventh in oil reserves, and nineteenth in natural gas reserves. Thus, coal is much more abundant than other fossil fuels in China; proven recoverable reserves of coal amount to 114.5 billion tons, more than 10 times the total reserves of currently proven oil and natural gas reserves, based on equivalent calorific values.

MIX OF PRIMARY-ENERGY SOURCES IN CHINA

In the past 20 years, China's economy has grown at a rate of more than 7 percent annually, and the consumption of primary energy has increased every

TABLE 1 Proven Recoverable Reserves of Fossil Fuels in the World, 2002

	World	China		United States	
		Quantity	Percentage	Quantity	Percentage
Coal (10^9t)	9,405	1,145	12.2	2,450	26.1
Oil (10^9t)	1,427	25	1.8	38	2.7
Natural gas (10^{12}m^3)	155.8	1.51	0.97	5.19	3.3

Source: CERS, 2003.

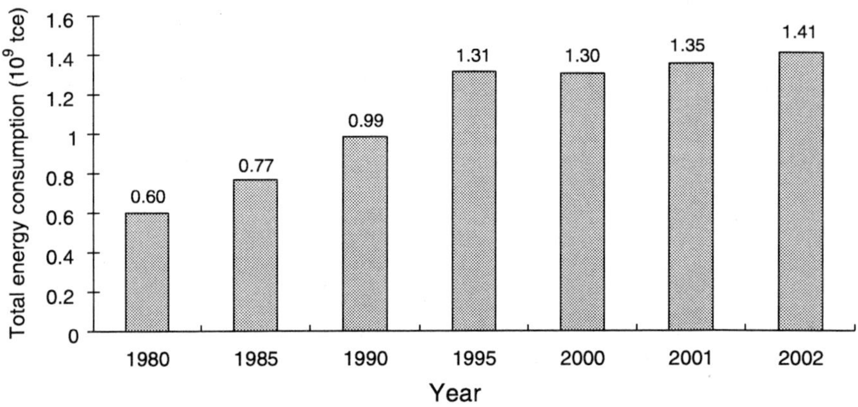

FIGURE 1 Energy consumption in China in the past 20 years. Source: National Bureau of Statistics, 2003.

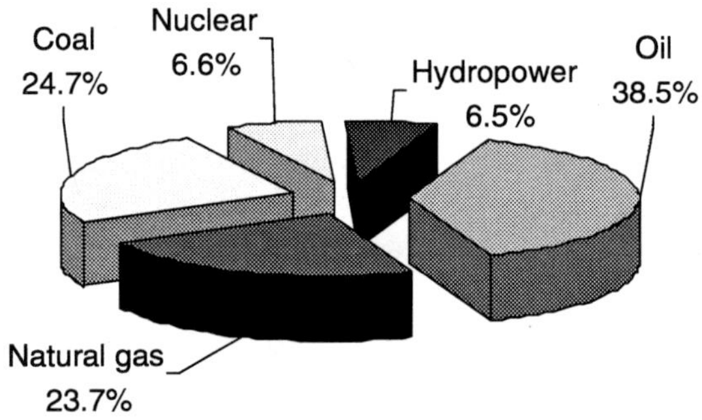

FIGURE 2 Consumption mix of primary energy in the world in 2001. Source: National Bureau of Statistics, 2003.

year (Figure 1). Fossil energy accounts for 86.9 percent of primary-energy consumption in the world, the highest proportion of which is from oil (38.5 percent). In China, however, the proportion of fossil fuels is 93.1 percent, the highest proportion of which is from coal (67 percent), 42.3 percent higher than the world average (National Bureau of Statistics, 2003). Consumption of oil and natural gas in China is 36.1 percent lower than in the rest of the world (Figures 2 and 3).

MIX OF END-USE-ENERGY SOURCES

Oil constitutes about 50 percent of the mix of end-use-energy sources consumed in the world. Natural gas and electric power constitute ~ 18 percent each, and coal constitutes 9 percent (Figure 4). In China, coal accounts for 44 percent,

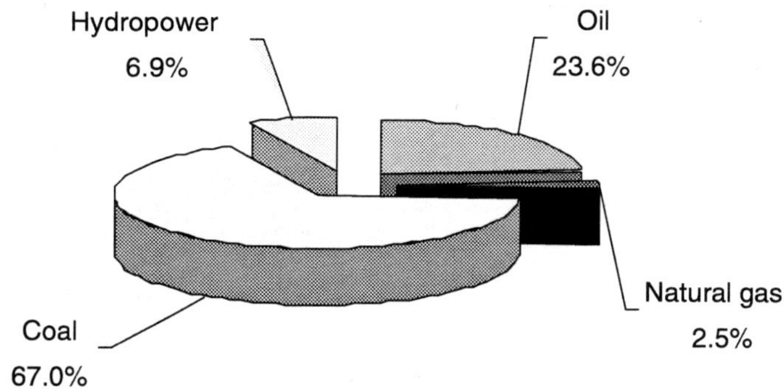

FIGURE 3 Consumption mix of primary energy in China in 2001. Source: National Bureau of Statistics, 2003.

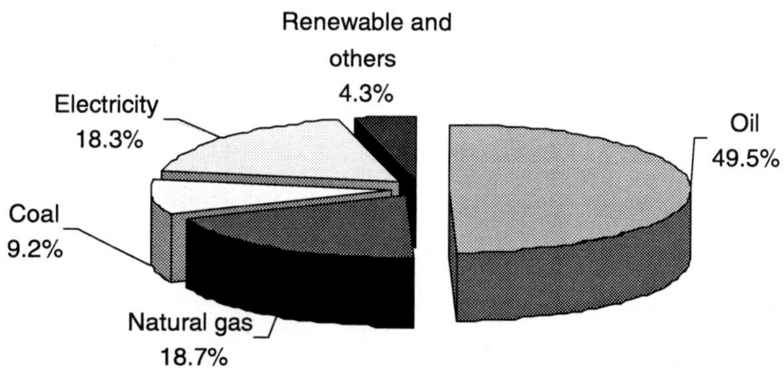

FIGURE 4 Consumption mix of end energy in the world in 2001. Source: National Bureau of Statistics, 2003.

oil 33 percent, and electric power 16 percent (Figure 5). Thus, the proportion of coal in the end-use-energy mix is 35 percent higher in China than in the rest of the world, and the proportion of oil and natural gas is 32 percent lower than in the rest of the world.

EMISSIONS

Particulates

In recent years, as a result of strengthened regulatory control of particulate emissions, 85.2 percent of power stations in China have installed electrostatic

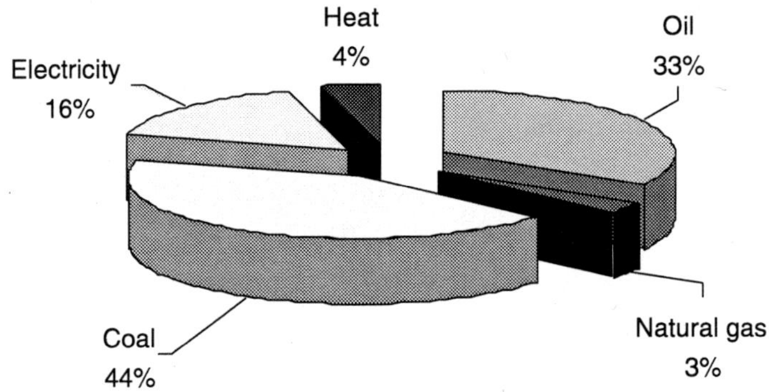

FIGURE 5 Consumption mix of end energy in China in 2001. Source: National Bureau of Statistics, 2003.

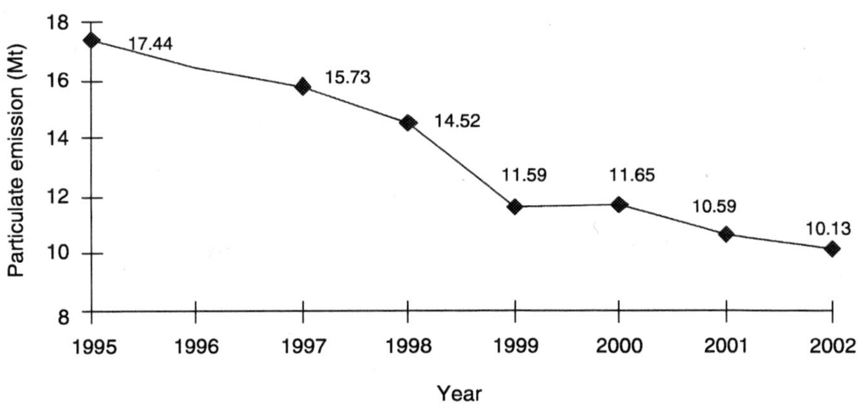

FIGURE 6 Particulate emissions, 1995–2002. Source: SEPA, 2003a.

precipitators. This has led to a marked decrease in total dust emissions (Figure 6). Fine particulates (PM_{10}), however, are still a major pollutant in urban areas, especially in cities in northern China (SEPA, 2003a).

Sulfur Dioxide

China has also strengthened its regulatory control of sulfur dioxide (SO_2) emissions. All new power plants that use coal with sulfur content of greater than 1 percent are now required to install desulfurization units; the rest are required to

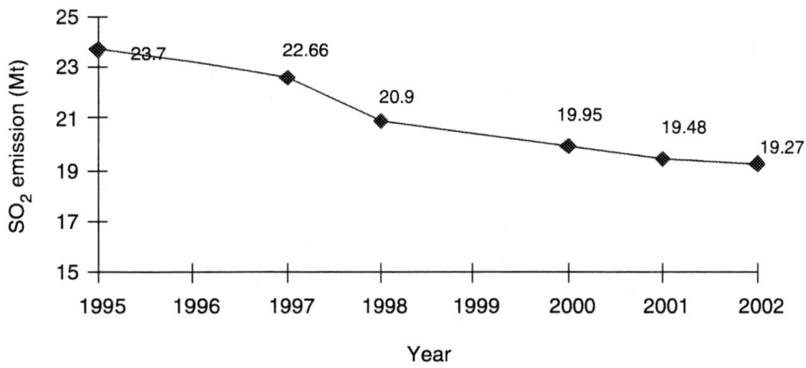

FIGURE 7 SO_2 emissions, 1995–2002. Source: SEPA, 2003.

TABLE 2 Annual Mean Concentrations of SO_2 Emissions in Chinese Cities (mg/m^3)

	1995	1996	1997	1998
National	0.080	0.079	0.066	0.056
Northern cities	0.081	0.083	0.072	–
Southern cities	0.080	0.076	0.060	–

Source: SEPA, 2003.

use low-sulfur coal. As a result, SO_2 emissions have been steadily reduced (Figure 7 and Table 2).

In 2001, 33.4 percent of the 341 cities monitored met or exceeded the national secondary standard for air quality (daily mean ambient concentrations of SO_2, nitrogen oxides (NO_x), and total suspended particles (TSP) of less than 0.15, 0.08, and 0.30 mg/m^3, respectively), and 33.2 percent failed to meet the national tertiary standard of air quality (daily mean concentrations of SO_2, NO_x, and TSP less than 0.25, 0.15, and 0.50 mg/m^3, respectively). Acid rain was evident in 161 cities (SEPA, 2003b).

Nitrogen Oxide Emissions in Cities

In 1999, the government initiated a Clean Vehicle Action Campaign, and all new power stations with an installed capacity of more than 300 megawatts (MW) were required to install low-NO_x burners (Yu et al., 2004a). In recent years, the annual average concentration of NO_x emissions has remained stable, and in some

TABLE 3 Annual Mean NO_x Emission Concentrations in Chinese Cities (mg/m^3)

	1995	1996	1997	1998
Mean values	0.047	0.047	0.045	0.037

Source: SEPA, 2003a.

places has even decreased slightly, even though the number of automobiles has increased significantly (Table 3).

PROBLEMS IN CHINESE CITIES

Low Energy Efficiency

The average energy efficiency in China is about 34 percent, about 10 percent lower than in European Union (EU) countries (Zhou and Wang, 2002). Low energy efficiency is apparent in mine-mouth power generation and domestic coal consumption.

Coal-Fired Power Generation

In 2000, the total production of electric energy in China was 1,350 terawatt hours (TWh). The average coal consumption for power generation was 390 gram coal equivalent per kilowatt hour (gce/kWh), 70 gce/kWh more than the world's most efficient power plants would consume (Editorial Board of China, 2003). Thus, in 2000, China consumed an extra 94 million tons of coal equivalent (Mtce) for power generation, the equivalent of about 132 million tons (Mt) of coal.

Industrial Boilers

There are about half a million industrial boilers in China with an average capacity of 2.5 tons per hour (t/h) and an average efficiency of 60 percent, 20 percent lower than the global average. Emissions from industrial boilers are higher than from power station boilers in almost half of Chinese cities. Thus, industrial boilers are the primary sources of pollution in many cities (Yu et al., 2004b).

Industrial Energy Consumption

The consumption per unit energy for industrial energy in China is 30 to 50 percent higher than in the world's most highly industrialized countries. In

2000, energy consumption per ton of crude steel in large and medium-sized iron and steel enterprises and the overall energy consumption of cement producers in China, were 18.6 percent and 54.4 percent higher, respectively, than in Japan (CERS, 2003). There is great potential for energy conservation in these areas.

Coal for Domestic Use

The efficiency of direct coal combustion for domestic use is very low, in some cases less than 20 percent. This is another area where great improvements can be made.

Slow Development and Dissemination of Clean-Coal Technologies

The energy technologies in many small and medium-sized facilities in China are out of date. In recent years, the state has promoted efficient, clean-combustion technology, clean-conversion technology, and new and renewable energy technology. Nevertheless, the development and use of these technologies has been limited.

Availability and Economics of Energy Sources

Table 4 shows energy prices in the United States in 1999, where the price ratio among coal, natural gas, gasoline, and electric power (equivalent calorific value) was approximately 1:3:8:16. The major factors that influence the selection of energy sources and clean energy technologies are energy availability, environment, and economics. One of the problems in China is the uneven distribution of energy sources. Seventy-seven percent of coal resources are in the north, in the area of the Qinling Mountains and the Huai River; 85 percent of oil resources are in the area north of the Yangtze River; 82.5 percent of water resources are in the western part of China, two-thirds of it concentrated in the southwest (Liu, 2002). However, energy is consumed mainly in eastern and central China. Because of

TABLE 4 Energy Price Forecast in the United States, 1999–2020 (average price for all customers in 1999 U.S. dollars per MBtu)

	1999	2005	2010	2015	2020
Coal	1.23	1.15	1.07	1.03	1.00
Natural gas	4.05	4.24	4.27	4.28	4.50
Liquid petroleum gas	8.84	8.84	8.88	8.58	8.26
Gasoline	9.45	10.64	10.93	10.75	10.68
Alcohol	14.42	19.12	19.00	19.24	19.36
Electric power	19.50	18.15	17.20	17.30	17.59

Source: CEPA, 2003.

the disproportional distribution of resources, coal has to be transported from north to south and from west to east in China. Currently, China is undertaking projects involving the transmission of electric power and gas from west to east.

Environmental requirements, energy availability, and energy prices vary greatly from city to city. Table 5 shows these variations for Beijing, Shanghai, and Chongqing. Beijing, the capital of China, which will be the host city for the

TABLE 5 Energy Availability and Economics in Beijing, Shanghai, and Chongqing

	Beijing	Shanghai	Chongqing
	Capital of China. Host city of Olympic Games in 2008. Depends on other provinces for most of its energy.	A major energy-consuming city. Depends mainly on other provinces and is remote from energy-producing sites.	Important energy-consuming city that produces high-sulfur coal, natural gas, and hydropower.
Available energy sources	Oil products, natural gas, coal, and some renewable energy.	Oil products, natural gas, and coal.	Oil products, natural gas, coal, hydropower, and methane.
Price of major energy sources			
Coal (5,000 kcal/kg)	About 260 yuan/t	About 380 yuan/t	120 yuan/t (indigenous high-sulfur coal). More than 300 yuan/t (from other provinces).
Natural gas (8,450 kcal/m^3)	1.8 yuan/m^3	1.8 yuan/m^3	1.2 yuan/m^3
Heavy oil (10,000 kcal/kg)	1,500 yuan/t	1,500 yuan/t	1,500 yuan/t
Light oil	2,200 yuan/t	2,200 yuan/t	2,200 yuan/t
Energy prices at equivalent calorific value	Natural gas > oils. Heavy oil > coal. (Natural gas nearly four times price of coal.)	Natural gas > oils. Heavy oil > coal. (Natural gas more than three times price of coal.)	Oils and heavy oil > natural gas > coal.

Source: China Energy Office, 2003; China Energy Online, 2003.

Olympic Games in 2008, will use natural gas as much as possible to guarantee good air quality for that event. The consumption of natural gas in Beijing is expected to increase from 1 billion m^3 in 2000 to 6 billion m^3 in 2008. Shanghai is expected to use a mix of energy resources. In addition to increasing the amount of natural gas it uses, Shanghai is devoting great efforts to the development of clean-coal technologies, such as supercritical power generation technology. Chongqing will use natural gas and hydropower as much as possible. At the same time, the proportion of electricity and heat converted from coal in Chongqing will increase. However, the proportion of coal in the mix of end-use energy will drop from 64.6 percent in 2000 to less than 50 percent in 2010 (Clean Energy Office, 2003).

CLEAN-ENERGY SOURCES

At present, less than half of the coal produced in China is used for power generation; in the United States, the proportion is more than 90 percent (Table 6); in EU countries it is 80.4 percent (CERS, 2002). From 1996 to 2000, the installed capacity of power-generating units in China increased from 236.54 GW to 319.32 GW, with an annual growth rate of about 8 percent. During those years, coal-fired power generation accounted for 78 percent of total power generation (National Bureau of Statistics, 2003).

Advanced Power-Generation Technology

Because the proportion of coal used for power generation in China is expected to increase significantly, the development of advanced power-generation technology is very important because it will reduce the amount of coal consumed, improve generation efficiency, improve the environment, and reduce emissions of carbon dioxide (CO_2). At present, the focus is on supercritical and ultra-supercritical generating units, which have been used successfully in other countries (Yu et al., 2004a). With the development of desulfurization and denitrogenation technologies and decreases in operating costs, this technology is now fairly competitive.

TABLE 6 Coal-Consumption Mix in China and the United States, 2000

	China	United States
Power generation	43.9 %	90.8 %
Coking	12.0 %	2.7 %
Industries and miscellaneous use	37.8 %	6.0 %
Domestic and commercial use	6.3 %	0.5 %

Source: CERS, 2002.

Technologies for Retrofitting Industrial Boilers

There are currently about half a million small and medium-sized industrial boilers in China that supply heat to industries and residences. Industrial boilers consume nearly one-third of coal production, and their total emissions equal, or even exceed, those of power-station boilers in some cities. Industrial boilers are not expected to be totally eliminated in the next 20 years in China. Therefore, it will be necessary to retrofit them (Yu and Chen, 2001).

Currently, "retrofitting" usually means substituting clean fuels for coal; using washed coal (i.e., coal with a sulfur content of less than 0.6 percent); or using sulfur-capture briquettes or coal-water mixtures (CWM). Research has shown that the economics of coal-fired industrial boilers varies greatly with different fuels. In northern China, the operating cost ratio among washed coal, sulfur-capture briquettes, CWM, natural gas, light oil, and heavy oil is 1: 1.2: 1.5: 3.1: 3.9: 2.3. Use of natural gas, light diesel, washed coal, briquettes, or CWM will all lower emissions to conform with the national environmental standard (CAE, 2001).

Low-sulfur coal provides the optimal ratio of economic benefits to environmental benefits. If all industrial boilers in China used low-sulfur coal, about 51 Mt of coal would be saved each year. Chinese cities must choose technologies and fuels for industrial boilers in accordance with environmental requirements, energy resources, and economic capacities (Yu et al., 2004a).

Combined Heat and Power Generation with Circulating Fluidized Boilers

Combined heat and power generation (CHP) has comprehensive benefits, like energy savings, lower emissions, better heating quality, and increased peak-power-shaving capacity. Cogeneration plants in cities could replace a great number of scattered heating boilers and industrial boilers. CHP would ensure the heating supply in winter, would be beneficial for industry, could supply electric power to some residents, would effectively control air pollution, and would improve energy efficiency.

Circulating fluidized bed (CFB) technology is developing rapidly in China. In the past 10 years, CFB boilers with capacities of up to 220 t/h have been commercialized. CFB boilers provide high-efficiency, low-pollution combustion and have the great advantage of operating with inferior fuels. In-bed desulfurization and low-temperature combustion effectively reduce emissions of SO_2 and NO_x, and CFB boilers used with power-generating units would guarantee clean CHP. Even though CFB technology requires a higher initial investment than other heating systems, the economic and environmental benefits can offset the expense. Calculations based on equivalent calorific values show that the operating cost ratio among CHP based on CFBs, gas-fired boilers, and oil-fired boilers is about 1: 1.7: 1.2 (Yu et al., 2004a).

ENERGY FOR COMMERCIAL AND DOMESTIC PURPOSES

Table 7 shows a comparison of the energy sources for domestic consumption in China and the United States. Natural gas and electricity are the primary sources of domestic energy generation in the United States, whereas in China, coal is the main energy source. With rapid economic development and increasingly stringent environmental requirements, cities in China are increasing the proportion of clean-energy sources for domestic and commercial use as their economic capabilities and the availability of energy sources allow.

In Beijing, Shanghai, and well developed coastal zones where environmental requirements are stringent, natural gas, liquefied petroleum gas (LPG), and liquefied natural gas (LNG) are the main energy sources being considered for commercial and domestic purposes. For heating, natural-gas-fired boilers, CHP generation based on coal-fired boilers, and central heating will replace dispersed coal-fired boilers. In less developed areas and areas that lack clean-energy sources like natural gas, coal-fired CHP generation, central heating, and advanced industrial boilers that burn low-sulfur coal will be used for heating. Briquettes and coal-saving stoves, which will be the main modes of energy consumption for commercial and household purposes, are also being developed (Yu et al., 2004a).

By the end of 2000, China had built 35.6 million square meters of energy-conserving buildings. But the technical standards for energy-saving buildings in China are about 50 years behind those of more highly industrialized countries. Therefore, compared with developed countries, the level of energy-saving technology in China is low. Energy consumption per unit of residential area for

TABLE 7 Domestic Energy-Consumption Mix in China and the United States, 1999

	China	United States
Consumption (Mtce)	106.4	368.1
Population (billion)	1.275	0.281
Mix (%)		
Natural gas	7.4	47.5
Liquid petroleum gas	14.2	4.5
Oils	3.5	9.4
Coal	51.4	0.4
Electric power	17.1	38.2
District heating	6.4	–
Total	100.0	100.0

Source: CERS, 2002.

heating in northern China is three times the consumption in developed countries with similar climates (Wang, 2003). Thus, energy-saving design standards for buildings, improved building designs, new building materials, improved thermal isolation in buildings, and green lighting and cold-accumulation air conditioning could be extremely beneficial for China.

China has comparatively abundant geothermal resources, which should be developed as quickly as possible for power generation, heating energy, industrial energy, and medical and other uses in cities where conditions are favorable. In 2000, the geothermal heating area covered 6 million square meters in Tianjin. Beijing already exploits 8.8 million square meters of geothermal water per annum, equivalent to 75,000 tons of coal. This has reduced dust emissions by 750 tons, SO_2 emissions by 1,300 tons, and CO_2 emissions by 34,000 tons per year, thus reducing air pollution and environmental damage (Clean Energy Office, 2003).

Municipal Traffic

In recent years, the number of vehicles in China has been increasing rapidly, causing severe tailpipe pollution in Chinese cities. NO_x, fine particulates, and dust emissions in large cities like Beijing, Shanghai, and Guangzhou exceed the national secondary standard. As the rate of urbanization increases, traffic density will also increase, and tailpipe pollution can be expected to worsen.

Bus Rapid Transit (BRT), a new transportation system, with costs comparable to those of an ordinary bus system and capacity roughly comparable to that of rail transport, was demonstrated in the city of Kunming during the 1999 International Horticultural Exposition. The Kunming Special Traffic Lane Demonstration Project was a joint project of Kunming and Zurich. The BRT system, which operates in a special traffic lane, features rapid boarding and de-boarding and a rapid ticket-purchasing system. Since 1999, the number of passengers transported daily has increased from 0.5 to 0.75 million, and BRT is now the hub of the municipal transit network. BRT has improved the flow of people and reduced the volume of traffic, thus decreasing tailpipe and noise pollution from vehicles (Municipal Planning Bureau of Kunming City, 2003).

THE CLEAN ENERGY ACTION PLAN

At the end of 2001, the Chinese government initiated the Clean Energy Action Plan to control air pollution caused by coal combustion. Eighteen pilot cities were chosen to promote clean-energy sources and clean-energy technologies. The first step in the plan was to assess the status of energy consumption and environmental pollution in the pilot cities. The results showed that in 2000, total consumption of primary energy was 207.6 Mtce, average consumption of

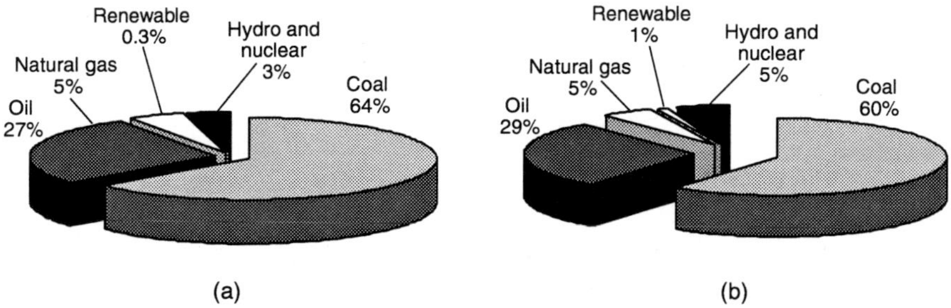

FIGURE 8 Primary energy consumption mix in 18 cities in (a) 2000 and (b) 2005 (estimated). Source: Clean Energy Office, 2003.

primary energy per city was 11.5 Mtce, and average coal consumption per city was 7.7 Mtce. The proportion of coal consumption in 13 of the 18 cities was more than 60 percent. The average oil consumption per city was 2.3 Mtce; oil consumption in seven cities was more than 20 percent. The average natural gas consumption per city was 2.7 billion cubic meters; gas consumption in only two cities was more than 10 percent.

Figure 8 shows the mix of primary energy consumed in the 18 cities in 2000 and 2005. In 2000, mean coal consumption was around 64 percent; oil and oil products accounted for 27 percent; and the total of natural gas, hydropower, nuclear power, and wind power was about 9 percent. It is expected that by 2005, as a result of the implementation of clean-energy sources and clean-energy technologies, the consumption of coal for primary energy will be reduced to 60 percent, oil and oil products will increase to 29 percent, natural gas will increase to 5 percent, and hydropower, nuclear power, and wind energy will be about 6 percent.

In 2000, the proportion of coal was more than 60 percent in six cities, in the range of 40 to 60 percent in six cities, and less than 40 percent in six cities; the national average was 38.9 percent. This very high consumption of coal contributed significantly to the atmospheric pollution in these cities. By 2005, the mix of end energy consumed will become cleaner, and air quality is expected to improve tremendously.

Air Pollution

The mean level of SO_2 emissions in the 18 cities was 140,000 tons, and the daily mean concentration of SO_2 emissions was 0.084 mg/m^3; both were within the national secondary standard. The mean TSP emission was 65,000 tons, and the daily mean concentration of TSP was 0.39 mg/m^3, exceeding the limit specified in the national secondary standard. Mean NO_x emissions were 44,000 tons, and the daily mean NO_x emission concentration was 0.05 mg/m^3, within the national secondary standard.

It is expected that by 2005 the mean SO_2 emission will be reduced to 110,000 tons in the 18 cities, and the daily mean concentration will be 0.062 mg/m^3; the daily mean TSP emission concentration will be 0.22 mg/m^3; NO_x emissions will basically remain unchanged. All emission levels will meet the national secondary standard.

CASE STUDY OF YINCHUAN MUNICIPALITY

Yinchuan is the capital of the Ningxia Hui Autonomous Region in western China. The total regional area is 3,500 km^2; the urban area is ~1,300 km^2. The total population is more than one million. The GDP value of the municipality in 2000 was 10.4 billion yuan. Mines that produce excellent quality coal are located on the periphery of the city. Natural gas from gas fields in Shaanxi, Gansu, and Ningxia provinces is transmitted by pipelines through Yinchuan City to eastern China. The mix of energy sources for Yinchuan in 2000 is shown in Table 8 and Figure 9. Yinchuan is also in the "high-quality luminous energy zone"; that is, the

TABLE 8 Primary Energy-Consumption Mix in Yinchuan in 2000

	Total consumption	Coal	Crude Oil	Natural Gas	Solar Energy
Consumption (1,000 tce)	2,686.9	1,576.5	757.5	343.6	8.9
Proportion (%)	100	58.67	28.2	12.79	0.34

Source: Clean Energy Office, 2003.

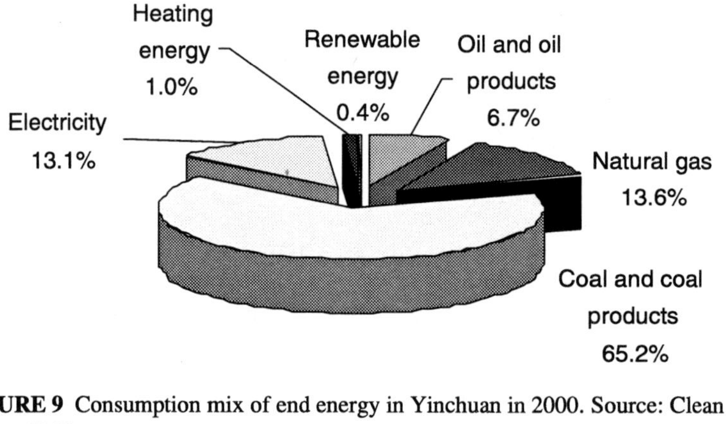

FIGURE 9 Consumption mix of end energy in Yinchuan in 2000. Source: Clean Energy Office, 2003.

TABLE 9 Air Pollution in Yinchuan ($\mu g/m^3$)

	1999	2000	2001
Mean annual TSP emission concentration	0.414	0.341	0.345
Mean annual SO_2 emission concentration	0.089	0.054	0.049
Mean annual NO_x emission concentration	0.044	0.036	0.032

Source: Clean Energy Office, 2003.

annual mean of hours of sunshine is more than 3,000, making solar energy a promising prospect.

TSP and SO_2 emissions have been declining gradually in Yinchuan in recent years (Table 9). Currently, the main air pollutant is TSP; emissions in 2000 were 0.7 times higher than the limit specified in the national secondary standard. Emissions of SO_2 and NO_x were below the national secondary standard. The main cause of severe pollution was coal combustion for heating during the winter, when concentrations of the major pollutants, such as TSP and SO_2, were high. From April to May and from November to December each year, strong winds bring blowing sand and dust, which causes seasonal high concentrations of TSP.

Clean Energy Options

Energy demand in Yinchuan is expected to be 3.7 Mtce in 2005 based on an average annual GDP growth rate of 9.5 percent and energy elastic coefficient of 0.31. The government's goals for environmental quality in Yinchuan in 2005 are shown in Table 10 and Figure 10. Table 11 shows projected options for the mix of energy sources in Yinchuan based on these goals:

- **Low option.** Do not change the mix, and do not employ clean-coal technologies. Under this scenario, SO_2 and particulate emissions would greatly exceed the emissions targets.

TABLE 10 Emission-Control Targets in Yinchuan for 2005 (1,000 tce)

	SO_2	Industrial Dusts
Actual value in 2000	26.3	16.7
Target value in 2005	21	17.8

Source: Clean Energy Office, 2003.

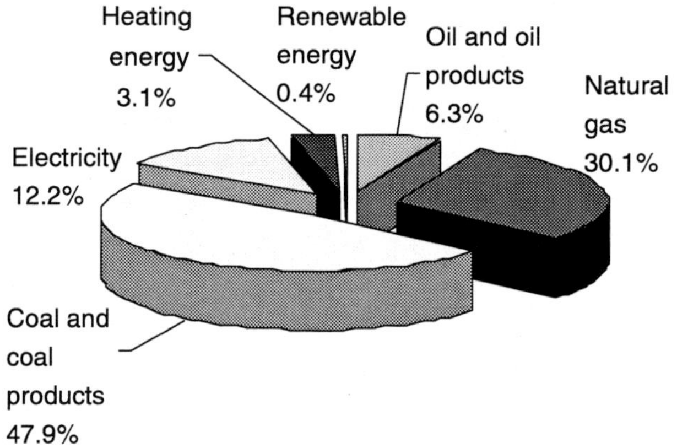

FIGURE 10 Consumption mix of end energy in Yinchuan in 2005. Source: Clean Energy Office, 2003.

TABLE 11 Options for Changing the Energy Consumption Mix in Yinchuan for 2005

	High Option	Medium Option	Low Option
Mix of End Energy (%)			
Coal	23.6	47.9	65.19
Natural gas	53.20	30.1	13.6
Oil	6.7	6.3	6.7
Heating energy	3.1	3.1	1.0
Electricity	13.1	12.2	13.1
Renewable energy	0.4	0.4	0.4
Air quality	In compliance with standard	In compliance with standard	Exceeds standard
SO_2 (Mt)	1.81	2.1	5.03
Particulate (mg/m^3)	0.33	0.86	1.25
Investment (billion yuan)	2.48	0.69	
Share of GDP in 2005	16.5%	4.6%	

Source: Clean Energy Office, 2003.

- **High option.** Replace coal with natural gas for all domestic heating. This would effectively improve air quality in 2005, and emissions would not exceed the national secondary standard. However, this option would require an investment of 2.48 billion yuan, or about 16.5 percent of GDP of Yinchuan City. In addition, annual operating costs would increase by 100 million yuan, which is unacceptable.

- **Medium option.** This option has been accepted for implementation. Clean energy will be substituted and clean energy technologies developed to meet the environmental targets for 2005. Six projects are planned: (1) substitution of natural gas for coal for domestic energy; (2) use of electricity instead of coal for some household cooking; (3) expansion of thermal power plants into CHP plants; (4) burning of low-sulfur coal instead of raw coal in industrial boilers; (5) equipping of industrial boilers above 6 t/h with desulfurization and flue-dust cleanup units; and (6) use of dual fuels (oil and gas) in automobiles.

Implementing the medium option will improve the end-energy sources substantially. Emissions of SO_2 will be reduced by 28,000 tons, and emissions of particulates will be reduced by 15,000 tons, which will satisfy both national and Ningxia environmental standards. The total investment will be 690 million yuan, or 4.6 percent of Yinchuan's GDP in 2005 (Clean Energy Office, 2003).

CONCLUSION

As the case study of Yinchuan municipality shows, meeting ideal environmental standards for Chinese cities is not practical in the short term. To improve air quality and promote the rational development of clean-energy sources and clean-energy technologies, each city will have to work out a practical, realistic, step-by-step, local energy plan based on many factors, including national environmental targets, the availability of indigenous energy resources, and the economic capacity of the city. This is the only way China can achieve coordinated development of energy, the environment, and the national economy.

REFERENCES

China Energy Online. 2003. Availability of Energy Sources. Available online at: *http://www.globalenergy.cn/*. In Chinese.

CERS (China Energy Research Society). 2002. Energy data for China and the world. Energy Policy Research 2002(1): 57.

CERS. 2003. Energy Data for China and the World. Energy Policy Research 2003(6): 76.

CAE (Chinese Academy of Engineering). 2001. Assessment of Rational Technical and Economic Routes for Coal-Fired Industrial Boilers. Beijing: Chinese Academy of Engineering.

Clean Energy Office. 2003. Clean Energy Plan. Available online at: *http://cct.org.cn/*. In Chinese.

Editorial Board of China. 2003. China Power Yearbook 2003. Beijing: China Power Press.

Liu, J. 2002. Study of China's Resources-Utilization Strategy. Beijing: China Agriculture Press.

Municipal Planning Bureau of Kunming City. 2003. Planning of Road Net System in Central Zones in Kunming. Available online at: *http://www.kmpg.gov.cn/ghj/index.asp*. In Chinese.

National Bureau of Statistics. 2003. China Statistical Yearbook, 2003. Beijing: China Statistics Press. Available online at: *http://www.stats.gov.cn/tjsj/*. In Chinese.

SEPA (State Environmental Protection Administration). 2003a. China Environment Yearbook, 2003. Beijing: China Environment Press.

SEPA. 2003b. Report on the State of the Environment in China. Available online at: *http://www.zhb.gov.cn/*. In Chinese.

Wang, G. 2003. Optimization of Energy Source Mix in Cities and Promotion of Energy Construction. Presented at the High-Level Energy Forum, Beijing, Nov. 16–17, 2003.

Yu, Z., and G. Chen. 2001. Burning washed coal: an effective way to reduce pollution from industrial boilers. Energy Conservation and Environment Protection 2001(5): 24–27.

Yu, Z., et al. 2004a. Clean Coal Technology Development and Application. Beijing: Chemical Industry Press.

Yu, Z., R. Zhou, and L.V. Xin. 2004b. Emission Reduction Measures and Assessments for Industrial Boilers in 18 Cities. Presented at the U.S.-China Industrial Boiler Workshop, Beijing, China, June 10–11, 2004.

Zhou, F., and Q. Wang. 2002. Fifty Years of Chinese Energy. Beijing: China Power Press.

Programs to Control Air Pollution and Acid Rain

SARATH K. GUTTIKUNDA and TODD M. JOHNSON
Environment Department
World Bank

FENG LIU
Energy and Mining Sector
World Bank

JITENDRA J. SHAH
East Asia Environment and Mining Sector
World Bank

Sulfur dioxide (SO_2) emissions from coal combustion are a primary contributor to acid rain and poor local air quality in China. Besides having adverse effects on human health, acid deposition has been recognized as an environmental threat to China's agricultural productivity. Acidic substances adversely affect aquatic systems, forests, monuments, and regional climates and alter the sensitivity of lakes, forests, soils, and ecosystems. In the long term, acids leach nutrients from the soil and diminish agricultural yields.

These effects are already being felt in the agricultural sector; an estimated 19 percent of the agricultural land in seven provinces (Jiangsu, Zhejiang, Anhui, Fujian, Hunan, Hubei, and Jiangxi) in southern China has been affected by SO_2 and acid rain. The average decrease in crop yield attributable to the combined effects of SO_2 and acid rain was 4.3 percent in the mid-1990s. Vegetable yield was reduced by 7.8 percent, wheat by 5.4 percent, soybeans by 5.7 percent, and cotton by 5.0 percent. In the same seven provinces, 4.2 percent of forests have been affected by acid deposition (Yang et al., 2002).

Other ecosystems are also beginning to suffer. A study of oak and pine trees affected by acid rain in both rural and urban areas of the Democratic Republic of Korea (North Korea) showed significant declines in growth rates since 1970 (Downing et al., 1997).

SO_2 emissions are also known to contribute significantly to fine particulate matter (PM) through formation of sulfate particles. Fine particulate compounds from sulphates and nitrates (formed by oxidation of emissions of sulfur and nitrogen oxides [NO_x], respectively) are often transported in the air over long distances. The health effects of particulates are strongly linked to particle size.

The constituents in small particulates tend to be chemically active. In parts of China, North Korea, and Thailand, sulfates are estimated to contribute significantly to the ambient particulate concentrations (PM_{10} and $PM_{2.5}$) (Guttikunda et al., 2003). The majority of Chinese cities have unhealthy levels of fine particulate concentration.[1]

The health effects of PM have been shown to be correlated with respiratory disease, which presently accounts for 18 percent of all deaths in China. A growing number of epidemiological studies have shown that fine particulates penetrate deep into the human respiratory tract, aggravating asthma, heart and lung disease, and general lung functions (e.g., Akbar and Kojima, 2003). Although ambient air quality has improved, estimates of the health effects of PM pollution in China in 1995 resulting from violations of ambient air-quality standards included 178,000 premature deaths, 346,000 registered hospital admissions, more than 6 million emergency room visits, and more than 75 million asthma attacks. The costs associated with these negative health effects were equivalent to more than 4 percent of China's GDP (Downing et al., 1997).

Coal is the principal source of energy and a primary source of air pollution in China, which is both the largest consumer and the largest producer of coal in the world. China's coal consumption in 2000 was 1.17 billion metric tons, or 25 percent of the world total (EIA, 2002). With the rapid economic growth in China in the last two decades, coal use has doubled, driven by fast-growing thermal electricity generation, as well as growing demand in the industrial and domestic sectors.

Cities in China are heavily polluted by SO_2 and PM emissions, primarily from the combustion of fossil fuels by both small domestic stoves and large industrial plants, including coal-fired power plants and boilers, ore smelters, oil refineries, etc. Smaller stationary combustion sources, such as space heaters, also contribute to the problem, especially in urban areas during the winter. Besides SO_2 and PM, emissions include NO_x, carbon monoxide (CO), carbon dioxide (CO_2), volatile organic compounds (VOCs), and other greenhouse gases (GHGs). GHG concentrations are likely to increase from all sectors as incomes and industrialization increase, further contributing to both local and global environmental concerns.

The growing demand for transportation and the rapid increase in the number of vehicles on the roads have led to an increase in air pollution, including sulfur emissions. However, recent emission inventories published by the State

[1]China does not currently report monitoring data for fine particulate pollution. Of 341 Chinese cities, nearly two-thirds have annual total suspended particle (TSP) concentration levels above the Class 2 national ambient air quality standard (200 micrograms per cubic meter [$\mu g/m^3$]), the maximum level acceptable for residential areas.

Environmental Protection Administration (SEPA) and other environmental organizations in China and abroad suggest that the transportation sector emits more NO_x, VOCs, CO, and CO_2 emissions than sulfur. Approximately 2 percent of the total sulfur emissions in China in 2001 was attributable to transportation, compared to about 86 percent from the industrial and power sectors combined (Streets et al., 2003). However, sulfur emissions from the transportation sector are not insignificant and should be included in future analyses.

One objective of the present study and the associated technical assistance project is to help localities in China address several questions related to the planning and implementation of regulations to control SO_2 emissions and acid rain:

- What are the environmental consequences for specific localities of different pollution control strategies in terms of human health effects, agricultural productivity, and other activities?
- What are the relative costs of different plans to reduce sulfur emissions?
- Will the proposed strategies enable localities to meet the environmental targets set by the central government?

REGULATION OF SULFUR DIOXIDE EMISSIONS IN CHINA

SO_2 emissions from coal combustion have long been a major contributor to ambient air pollution in Chinese cities and are the primary cause of acidic precipitation in ecologically sensitive areas and much of China's most fertile land areas. By 1996, when China's coal consumption reached historic highs, ambient SO_2 pollution had become severe and widespread in major cities. Of the 90 cities with reported data, the median annual SO_2 concentration level was 60 micrograms per cubic meter ($\mu g/m^3$); the highest concentration was 418 $\mu g/m^3$, compared with the World Health Organization (WHO) guideline value of 50 $\mu g/m^3$.

Acid rain, defined as precipitation with a pH value lower than 5.6, had expanded from a few pockets in southwestern China in the mid-1980s to about 30 percent of the country's land area by the mid-1990s. Figure 1 shows the critical loads calculated in the RAINS-ASIA 7.5 model for China at 10 percentile (i.e., about 90 percent of the ecosystem in each grid will have a low risk of being damaged if the acid deposition does not exceed this 10-percentile critical load). A critical load is an estimate of the maximum allowable input of acid deposition that will not adversely affect growth or otherwise damage ecosystems. This report has been prepared for a 10-percentile critical load, but ecosystems in southern China, which are very susceptible to soil acidity, have lower critical loads.

In 1998, China adopted national legislation to limit ambient SO_2 pollution and halt the increase of acid rain. The program became known as the "two control zones (TCZs) plan," because of its geographical coverage of: (1) cities with high ambient levels of SO_2 that are subject to ambient concentration

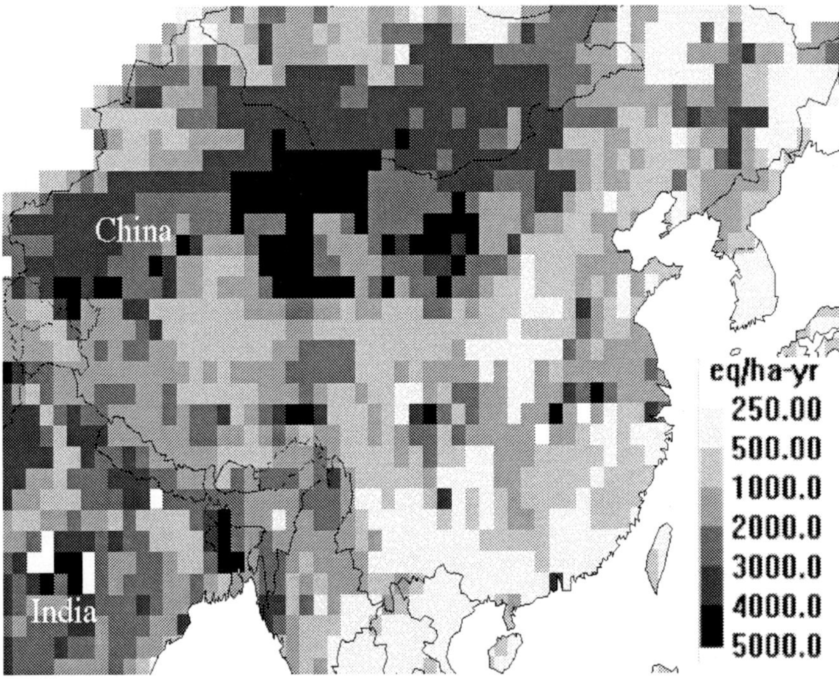

FIGURE 1 Critical loads for acid deposition at 10 percentile, in equivalent hectacres per year (eq/ha-year). Source: Downing et al., 1997.

compliance requirements; and (2) regions with serious acidification problems that are required to reduce the incidence of acid rain through reductions in SO_2 emissions (Figure 2) (Pu et al., 2000; SEPA, 2002).

Targets in the National Tenth Five-Year (2001–2005) Plan for Environmental Protection, included the following stipulations for 2005:

- Annual sulfur emissions in the TCZs must be reduced from their 2000 levels by 20 percent.
- Annual ambient SO_2 concentration levels of 31 noncompliant cities must meet the national standard for residential areas.

With the passage of the TCZ legislation, the Chinese government took an unprecedented step toward controlling sulfur emissions. By the late 1990s, ambient SO_2 concentrations in many densely populated urban areas were exceedingly high and harmful, and many incidences of acid rain had been documented in China's principal agricultural areas, including Sichuan Province. Backed by studies and expert opinion from leading Chinese universities and research

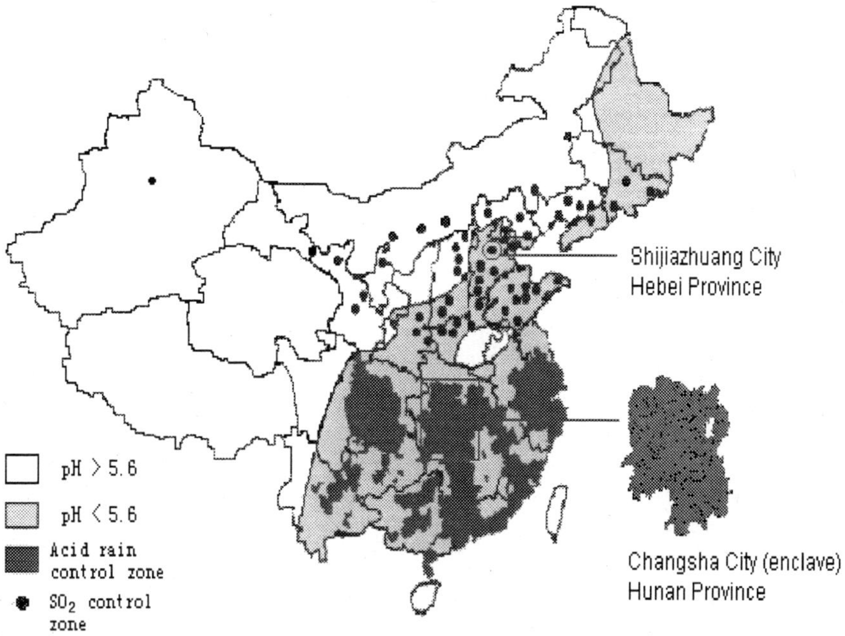

FIGURE 2 Two control zones in China with case study locations. Source: SEPA, 2002.

institutions, SEPA helped win approval of the sulfur-control legislation. At the time, most of the evidence of human health effects and acid rain damage was anecdotal, and there was no systematic assessment of the level or extent of the impact of ambient SO_2 levels and acid rain. Nevertheless, SEPA convinced the government of the importance of controlling SO_2 emissions, using evidence that reportedly included future scenarios of human health impacts and damage from acid rain on manmade structures, forests and other ecosystems, bodies of water, and especially agricultural production. The government was probably also influenced by international concerns about acid rain and China's growing contribution to regional and global SO_2 emissions.

Considering that SO_2 emissions have not been regulated in the past, the TCZ targets will be difficult to meet. China is the first developing country to regulate SO_2 emissions aggressively and on a large scale; the only parallels are sulfur-control legislation and control measures that have been adopted in Europe and North America. The cities and provinces in China affected by the TCZ legislation were required to submit detailed implementation plans to SEPA for controlling SO_2 by January 2003.

NATIONAL GOALS

National long-term goals are concerned mainly with SO_2 pollution generated by burning coal. The widely adopted pollution-control approaches fall into three categories: fuel switching; sulfur removal; and flue gas desulfurization.

Fuel Switching

Fuel switching usually involves the use of low-sulfur coal or sulfur-fixed briquettes at the lower cost end, and gaseous fuels, such as liquefied petroleum gas (LPG) and natural gas, at the higher cost end. These options represent a wide spectrum of costs and are subject to the constraints of local access to low-sulfur fuels. Some Chinese cities have also encouraged the use of electricity for cooking, water heating, and even space heating in recent years. However, if the electricity comes from coal-fired power plants with no controls on SO_2 emissions, this could simply dilute local pollution.

Sulfur Removal during Coal Combustion

Sulfur can be removed during coal combustion (in situ sulfur removal) in industrial and utility boilers by using sorbent-injection techniques or fluidized bed combustion (FBC) technology. The former usually involves injecting dry sorbent (either calcium-based or sodium-based) into the furnace of a boiler. The latter involves firing a suspended fine mixture of coal and sorbent (such as lime). Even though the basic sorbent-injection technique is easy to set up and operate and relatively inexpensive, it is rarely used in China, in part because of a lack of experience with the technology. Non-pressurized FBC boilers are beginning to penetrate the market, especially in the large boiler segment (i.e., 70-ton steam per hour or larger), as domestic manufacturers master the technology.

Flue Gas Desulfurization

Flue gas desulfurization (FGD) is most cost effective in coal-fired power plants. The most widely used FGD technology, wet scrubbers, uses gas/liquid reactions to remove sulfur from flue gas. A cheaper alternative, spray dry scrubbers, is usually used for small utility boilers or older plants. Both technologies have been demonstrated in China and, based on the large proposed investments in the Tenth Five-Year Plan, both appear to be in demand. So far, because of high costs and rigid utility pricing regulations, only a few facilities are operating with FGD technology.

Progress to Date

Many advanced coal utilization technologies, such as integrated gasification and combined cycle (IGCC) power generation, are highly efficient for sulfur removal but are still experimental and costly. Investments in energy efficiency can reduce the demand for coal and associated sulfur emissions, but they must be justified by the economics of energy savings; regulations on sulfur emissions could provide added incentives for such investment. The suitability of control measures at any given locality depends on the availability and cost of low-sulfur fuels, the cost of control technologies, the condition of existing facilities, and local (or even site-specific) emission-control targets.

Control of sulfur pollution became a major regulatory pursuit in China with the TCZ plan, which has prompted many coal-fired power plants to switch to low-sulfur coal. However, few power plants or industrial coal users have adopted specific sulfur-emission control technologies. Broad-based urban residential and commercial fuel-switching programs, which began more than a decade ago, have significantly reduced ambient SO_2 pollution, among other benefits. Thus, fuel switching is the only common measure for mitigating sulfur emissions.

CHALLENGES

Controlling sulfur pollution in China is more difficult than in North America or Europe for several reasons. First, China's economy is largely dependent on coal, and the demand is expected to increase over the next 20 years. Regulating SO_2 emissions will affect far more than coal-fired power plants in China (the electric power sector accounts for less than 50 percent of national coal consumption). About half of the population relies on coal-fired devices for space heating, and more than 400,000 small to medium-sized coal-fired boilers are used in industry and commerce. Controlling emissions from such a large number of dispersed users, especially where there are few available and affordable control measures, will be extremely difficult.

Second, capital for investing in environmental control is scarce in China. For both national and local governments, policy decisions involve not only balancing GDP growth and environmental protection, but also stretching resources to address concerns about air, water, solid waste, and natural resources. Until the mid-1990s, there were few regulatory or financial incentives for industry (including the electric power industry) to invest in sulfur-emissions abatement.

Finally, institutional capacity for managing air pollution in China is underdeveloped, and most local environment agencies do not have sufficient capacity to monitor and regulate sulfur emissions. The development and implementation of a permit system for large emitters of SO_2 are still at an early phase, and the regulation of numerous small coal users, although important, is very difficult. The lack of institutional capacity also limits urban-scale micro-analysis of sulfur-control measures and regulations.

CHINA'S NATIONAL SULFUR-CONTROL PROGRAM

Most localities in China lack the tools and assessment capabilities for determining the relative benefits of emission reductions and, therefore, have typically focused on quantifying reductions in total emissions, without regard to where those emissions originated or where they ended up. Unlike total emissions, however, pollution impacts from SO_2 are closely correlated with the spatial distribution of ambient concentrations and the incidence of acid rain. Thus, it is important to understand the dynamics of concentrations and geographic locations when planning abatement strategies.

Similarly, analyses of the costs of pollution control have typically focused narrowly on up-front capital costs of implementing measures to reduce emissions, without taking into account operating costs or the multiple benefits of some pollution-control measures (e.g., the use of natural gas). A better understanding of the relationship between emissions and impacts—the decrease in the cost of damage rather than the amount of emissions reduced—can also clarify the benefits and costs of emission controls and enable cities and regions to make better decisions about allocating scarce funds for environmental improvement.

This study, funded the by the World Bank Energy Sector Management Assistance Program (ESMAP), analyzes China's national sulfur-control program by looking at local implementation plans and actions for reducing sulfur emissions in two municipalities—Shijiazhuang and Changsha. The city of Shijiazhuang in Hebei Province, a northern Chinese city, was chosen for a case study on ambient SO_2 pollution control; the urban region of Changsha, Xiangtan, and Zhuzhou (CXZ) in Hunan Province was chosen to represent a southern area that has high levels of acid rain.

The case studies provide specific local lessons that can inform China's national sulfur-control policy and provide guidance on the effectiveness and consequences of measures for meeting national targets. Follow-up meetings with SEPA will be held to discuss the strengths and weaknesses of the sulfur-control policy in terms of reducing ambient air pollution and reducing acid deposition in areas covered by the TCZ policy.

Shijiazhuang City, Hebei Province

Shijiazhuang city, the capital of Hebei Province, is 275 km southwest of Beijing. The city has a population of about 1.6 million in an area of 254 km^2. Of the 110 km^2 of established (built-up) areas, 27 percent is residential, and 23 percent is industrial. The jurisdictional area of Shijiazhuang Municipality, which is much larger than the city, includes six districts and 17 counties; the municipality borders Shanxi Province and covers an area of 16,000 km^2, 56 percent of which is mountainous. The municipality as a whole has about nine million residents and is largely rural.

The sources of sulfur pollution in Shijiazhuang are largely central heating boilers and large point sources, both inside and outside the city limits. The highest annual average SO_2 concentrations reach 180 $\mu g/m^3$, and the health effects alone are equivalent to about 10 percent of local GDP. More than 90 percent of planned reductions in sulfur emissions through 2005 would come from fuel substitution—low-sulfur coal for industrial and power-sector boilers and natural gas for domestic cooking and heating and small industrial boilers (Table 1).

Even if Shijiazhung meets the proposed sulfur-control targets for the Tenth Five-Year Plan, it is likely to fall short of the ambient pollution standards

TABLE 1 Planned SO_2 Control Measures in Shijiazhuang, 2001–2005

Projects	Projected SO_2 Reduction (ktons/yr)	Estimated Investment (million yuan)	Measures
Supply of low-sulfur coal	19	225	1. Limit the sulfur content of coal sold in the city to less than 1 percent. 2. Promote the importation of low-sulfur coal from neighboring Shanxi Province.
Substitution of natural gas for coal	13	513	1. "Gasify" some central heating in the downtown area. 2. Increase the share of gaseous fuel in the fuel mix from the current 3 percent to at least 10 percent.
Small coal-fired boilers and kilns	3	100	1. Dispose of all coal-fired boilers with a capacity of less than 1 ton or require that they use low-sulfur anthracite or gas. 2. Replace 700 small boilers used for winter heating with district heating systems. 3. Convert 50 dispersed coal-fired heating boilers to electric or gas boilers.
Desulfurization	1	150	1. Apply water-screen desulfurization and introduce circulating fluidized-bed combustion methodology for thermal power plants.

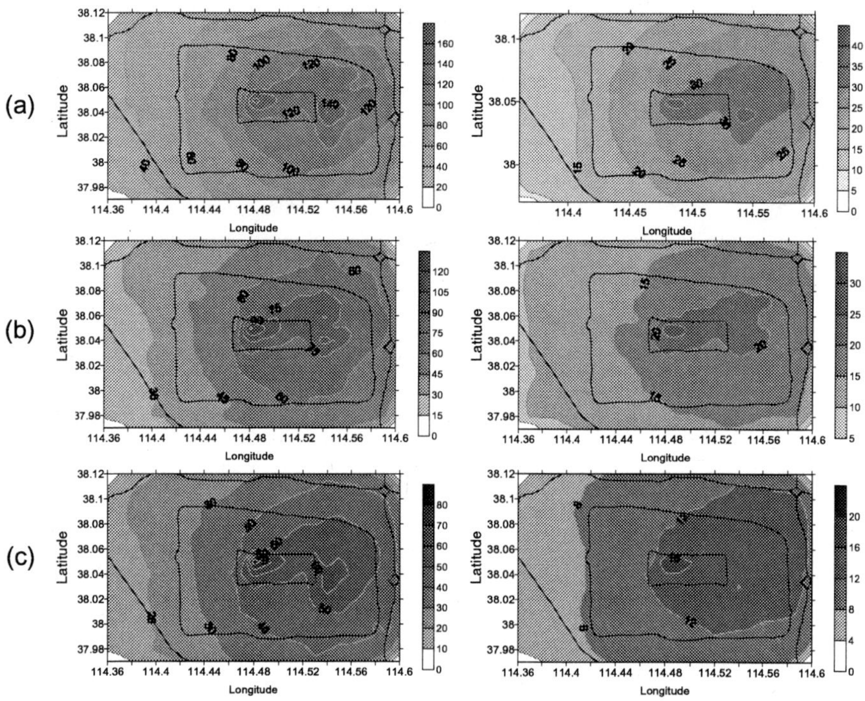

FIGURE 3 SO_2 and SO_4 concentrations (in μg/m³) in Shijiazhuang in 2000. 2a. Estimated SO_2 and SO_4 concentrations in Shijiazhuang. 2b. The base-case scenario (Scenario I) for 2005. 2c. Scenario II for 2005. Source: Downing et al., 1997.

required by SEPA. Shijiazhuang is an interesting example of both the potential and the constraints of using fuel switching as the main strategy for controlling ambient SO_2 pollution. The low-sulfur coal and natural gas options account for 53 and 36 percent, respectively, of planned reductions between 2001 and 2005. Even though the Tenth Five-Year Plan actions (Scenario I) are not sufficient for Shijiazhuang to achieve full compliance with the Class 2 ambient SO_2 standard, switching to low-sulfur coal at large point sources (Scenario II) would essentially bring Shijiazhuang into full compliance.

Figure 3 shows estimated SO_2 and SO_4 concentrations under baseline and control scenarios.[2] Both scenarios are based on the implicit assumption that coal

[2]More detailed analyses of the scenario emissions and modeling results can be found elsewhere (World Bank, 2003); the scenario analyses are based on other work (Li, 1998a; ECON, 2002).

consumption will not increase, at least between 2001 and 2005. This may be optimistic, unless more of the fuel supply comes from natural gas.

Capping sulfur emissions from space-heating boilers in houses and from power plants in and near the city proper will be essential for continuous compliance in Shijiazhuang. If growing demand for space-heating leads to an increase in coal consumption, further emission reductions at these sources will be necessary. The current strategy of consolidating small coal-fired central-heating systems into large district heating systems would facilitate emission control and compliance monitoring. In addition, most of the consolidation and expansion of space-heating capacity will use combined heat and power facilities, which will make it more economical to invest in large district heating systems in a relatively mild climate (the official heating season lasts four months, from November 15 to March 15, with an average outdoor temperature in the coldest month, January, of –4.6°C to –2.7°C).

Alternatively, if the natural gas supply increases and the price of natural gas becomes more competitive, distributed gas-fired space heating could become a practical option. The economic and financial implications of large district heating systems and distributed gas heating should be compared to provide guidance to the local government in making investments in infrastructure. In addition, the construction of more energy-efficient buildings and the introduction of consumption-based pricing and billing would significantly reduce future demand for heating fuels. This cross-sector policy would clearly help Shijiazhuang control air pollution.

Shijiazhuang should not add new coal-fired power plants in its vicinity. Strict environmental reviews of power projects will be necessary to ensure the consolidation of heating-boiler houses.

Finally, the emission abatement strategy adopted by Shijiazhuang will succeed only with strong policy and regulatory support, especially in the Tenth Five-Year Plan period. Extensive monitoring and enforcement of compliance will be necessary to ensure that low-sulfur coal is used by the many heating and industrial boiler houses, because the local coal market is completely decentralized and coal transactions are difficult to track. This and other regulatory issues are discussed in more detail in the final section of this paper.

Greater Changsha Including Xiangtan and Zhuzhou (CXZ), Hunan Province

This tri-city region is anchored by Changsha, the capital of Hunan Province; Xiangtan and Zhuzhou, two medium-sized industrial cities, are located at the lower points of the triangle. The CXZ region covers an area of about 4,500 km^2 and has the highest density of population in Hunan Province. In 1999, the population reached 12.3 million, about 19 percent of the population of Hunan. In the same year, the GDP of the area reached 108.3 billion yuan with an industrial output value of 58.6 billion yuan, accounting for 32 percent and 41 percent, respectively, of the provincial total.

Sulfur deposition in the greater Changsha region, a hot spot of acid rain in China, is dominated by emissions from large industrial smelters and power plants. SO_2 pollution is also high around the large power plants and smelters, with annual average ambient concentration levels exceeding 300 $\mu g/m^3$. Large areas, including rice paddies, fields planted with vegetables, and forests, have been damaged by acid rain. Claims of crop losses by peasants have been settled for millions of yuan.[3]

Planned reductions in sulfur emissions are predicated on a large increase in the use of natural gas and low-sulfur coal for industrial boilers. A significant amount of the reduction would be achieved through industrial renovations and the installation of pollution-control equipment at a single smelter in Zhuzhou (Table 2). Figure 4 shows a comparison of various options for reducing sulfur emissions in the tri-city area.

Figure 5 shows the sulfur wet-deposition option under two scenarios (for 2000 and 2005). Because acid deposition is a regional phenomenon, the results include sulfur deposition caused by sources outside Hunan Province, calculated using the RAINS-ASIA model. Total wet deposition in the region ranged from 0.5 g-S/m^2/yr for emission sources (mostly background deposition levels) to 5.0 g-S/m^2/yr near the Zhuzhou smelter. SO_2 concentrations were also high around the LPS locations, especially around the Zhuzhou smelter, with highs of 324 $\mu g/m^3$ (annual average). More detailed analyses of the scenario emissions and modeling results can be found elsewhere (World Bank, 2003).

The sulfur-control measures proposed in the Tenth Five-Year Plan are in line with emission-reduction targets set by SEPA for the CXZ region. However, even if they are met, the environmental costs of acid rain in the region would remain high. Sulfur-control targets are threatened in the short to medium term by the social and economic constraints of shutting down local coal mines that produce high-sulfur coal and by the planned construction of new coal-fired electricity-generating capacity near the city.

Although acid rain is the major air pollution problem in the CXZ region, ambient SO_2 pollution and fine particulate pollution are also critical issues. Because the primary cause of all of these problems is the burning of high-sulfur coal, abatement measures for acid rain will also address ambient air quality concerns. The proposed strategies for controlling sulfur emissions in the Tenth Five-Year Plan address the major emission sources, but are also formulated not to cause drastic cutbacks in the consumption of locally produced high-sulfur coal, at least in the near term.

[3]According to local EPB officials, many farmers have sued industries for crop loss in recent years. Local EPBs often act as mediators to resolve these cases by estimating crop losses based on historical harvests (discussion notes during the World Bank Mission, July 2001).

TABLE 2 Major SO_2 Control Measures Planned for the CXZ Region, 2001–2005

Projects	Projected SO_2 Reduction (tons/yr)	Estimated Investment (million yuan)	Measures
Supply of low-sulfur coal	8,400	330	1. Construction of coal-processing centers in Changsha and Zhuzhou.
Increasing the use of gaseous fuels	31,600	1,635	1. Increase in natural gas supply to Changsha, Xiangtan, and Zhuzhou. 2. Distribution of piped mixture of LPG and air in Changsha. 3. Recovery of fugitive coal-gas in Xiangtan.
Emission control of coal-fired boilers and kilns	15,000	325	1. Disposal of small coal-fired boilers in Changsha. 2. Introduction of wet precipitators for medium-sized boilers in Changsha. 3. Control of fugitive emissions in Zhuzhou. 4. Control of boiler smoke in Xiangtan. 5. Provision of centralized steam supply in Changsha and Xiangtan.
LPS Desulfurization	22,600	250	1. Application of desulfurization techniques at Zhuzhou power plant and Xiangtan fertilizer plants. 2. Introduction of TOPSOE method to convert SO_2 to commercial sulfuric acid at Zhuzhou smelter—expected to be operational in 2003.[a]

[a] HALDOR TOPSOE, "The Catalyst and Technology Company," specializes in wet gas sulfuric acid (WSA) processes. Source: Downing et al., 1997.

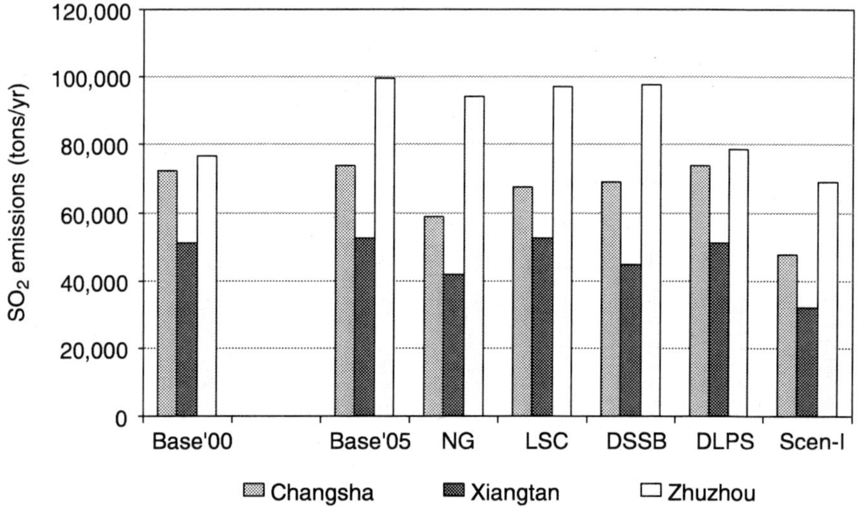

FIGURE 4 Comparison of estimated emissions under the baseline scenario in 2000 (Base '00) and Scenario I in 2005 (Scen-I). Base '05 is the no-controls scenario for 2005. Emissions under scenarios for natural gas (NG), low-sulfur coal (LSC), desulfurization for small-scale boilers (DSSB), and desulfurization for large point sources (DLPS) are also presented.

If implemented successfully, the plan would meet SEPA's requirements for reducing regional sulfur emissions and would significantly reduce damage caused by acid rain. Further reductions of sulfur emissions after the Tenth Five-Year Plan would have to be focused on area sources, mostly small and medium-sized boilers that continue to burn high-sulfur coal, coal-fired power plants, and dirty smelters in the region. Success will depend on strong policy support for enforcing current regulations and substantive government support in arranging for, or assisting in, the financing of key projects, such as increasing the natural gas supply, introducing LPS desulfurization, and ensuring a supply of cleaner coal.

Two issues of particular interest to the CXZ region are highlighted below: (1) the abatement strategy for the industrial sector; and (2) the abatement strategy for the power sector. (Issues that apply to both the CXZ region and Shijiazhuang are discussed in the concluding section of the paper.)

Abatement Strategy for the Industrial Sector

There appear to be two distinct strategies that would lead to different investment decisions and have different socioeconomic implications. The strategy implied in the current Five-Year Plan is focused on phasing out high-sulfur coal

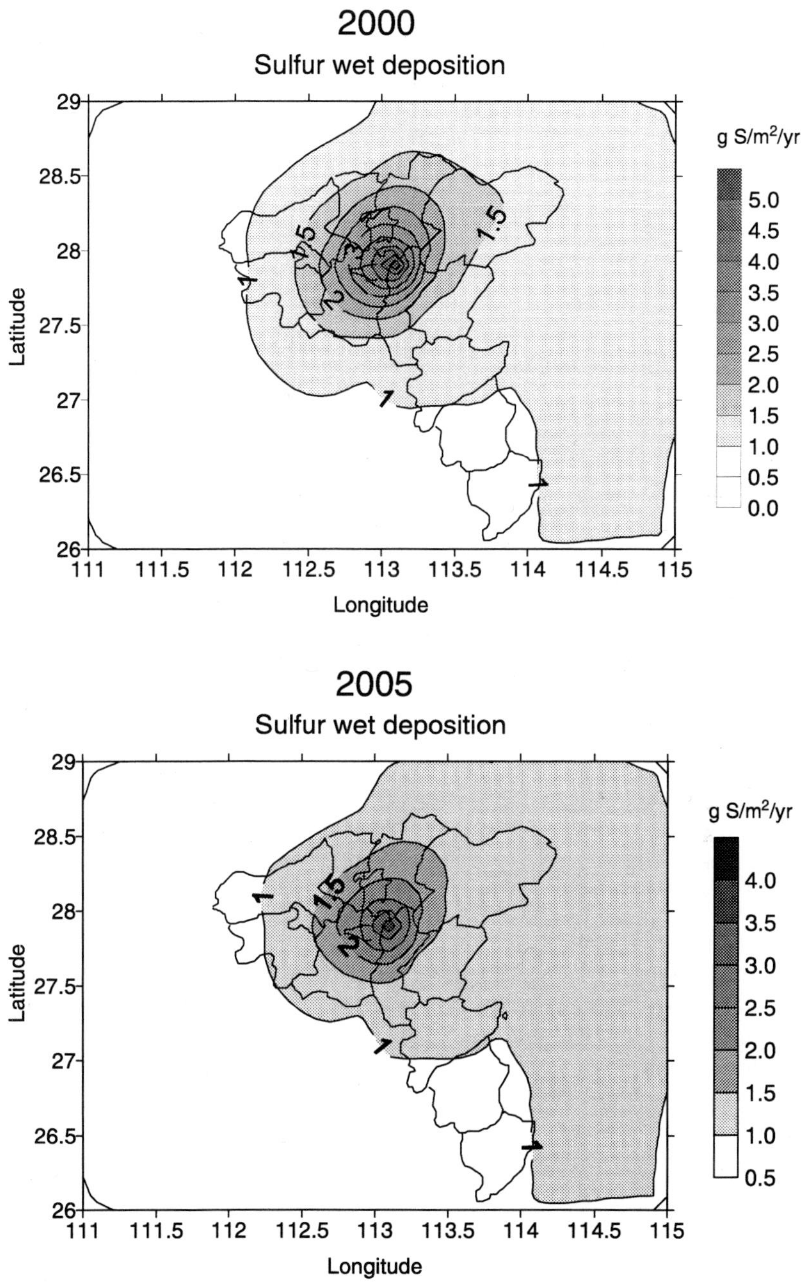

FIGURE 5 Estimates of sulfur wet deposition. 5a. Baseline scenario 2000. 5b. Scenario I 2005. Source: Downing et al., 1997.

consumption, either by increasing the use of low-sulfur coal from northern China or by substantially reducing the sulfur content of local (Hunan) coal by blending, washing, and briquetting, perhaps in combination with additional emission-control measures. The policy objective appears to be the development of a market for cleaner coal among industrial coal users and large commercial operators. The obvious advantage of this strategy is that it requires minimal technical adjustments and low up-front capital investments by end users. However, because this strategy focuses on centralizing and streamlining the local coal supply and distribution system, implementation may require substantial reorganizing by the government and possibly large public financing.

An alternative strategy could be the introduction of FBC boilers that burn high-sulfur coal directly but can also reduce sulfur emissions. FBC boilers burn coal more efficiently than conventional fixed-bed or chain-grate boilers. The policy objective of this strategy is to deploy a cleaner coal-burning technology among industrial coal users and large commercial operators. The most appealing aspect of this strategy is that it would help sustain the local coal industry while keeping sulfur emissions in check. However, FBC technology is unfamiliar to local boiler operators and is relatively expensive and sophisticated. Thus, FBC boilers are likely to be deployed over a long period of time as aging boilers are replaced. Most of the current stock of industrial boilers is likely to turn over in 10 to 15 years, providing a window of opportunity for FBC boilers.[4]

The differences between the two strategies are apparent. Large-scale deployment of FBC boilers would make investments in the supply of cleaner coal unattractive, and vice versa. The socioeconomic and technical merits of both strategies require further investigation to justify government support for one or the other.

Abatement Strategy for the Power Sector

In response to large increases in the demand for electricity in the CXZ region and Hunan Province as a whole, major power projects are already under construction or in preparation. According to the Hunan Province Environmental Protection Bureau (EPB), new power plants will be the new sources of sulfur emissions. Hunan is part of the Central China Power Grid that also covers Hubei, Henan, and Jianxi provinces. The Central China Grid will share the power from the Three Georges Dam (18.2 GW and 84.7 TWh) with two other grids, and perhaps part of the compensation capacity for the dam's high seasonal

[4]The World Bank Global Environment Facility (GEF) China High-Efficiency Industrial Boiler Project has FBC technology components. This project is nearing completion, and the technologies developed are just entering the commercialization phase.

variation of power generation. Thus, planning for the CXZ region is closely linked to planning for Hunan Province and the Central China Grid as a whole.[5]

The current policy implies that the region is inclined to adopt a build-and-control strategy, meaning erecting power plants in the region and investing in FGD or importing low-sulfur coal, whichever is sufficient for the CXZ area to comply with environmental regulations. The alternative would be to adopt a buy-and-avoid strategy, meaning importing electricity and avoiding the construction of new power plants in the region altogether. Unlike the industrial sector, the CXZ area itself may have little influence on decisions related to the power sector, which tend to be centralized. It may also be politically appealing to local governments to support the construction of new power plants. However, a previous plan for a large coal-fired power plant in the Changsha area was canceled in part because of concerns about pollution.

From an environmental point of view, the CXZ area, which is already a national hot spot for acid rain damage, may not be an ideal place for new coal-fired power plants, which would add to human health and agricultural damage. Economically speaking, electricity from a new coal-fired power plant in the CXZ area may not be cheaper than buying hydropower from Hubei or thermal power from Henan, both of which are in the same regional power grid. In addition, a large coal-fired power plant could easily eat into new supplies (or transport capacity) of low-sulfur coal to the region, affecting allocations to the non-power sector. A careful review of the electric power development plan in the CXZ area, and perhaps in Hunan as a whole, may be necessary to ensure that it meets the overall long-term economic and sulfur-control goals.

FINDINGS AND LESSONS LEARNED

These case studies and information gathered from other experience with sulfur-emission control in China support the following findings.

Finding 1. There is a clear divide between the northern and southern cities and regions in China in terms of the impact of SO_2 emissions and in terms of potential solutions.

Acid rain is a mostly southern phenomenon; high ambient SO_2 levels are more prevalent in northern cities where winter space heating exacerbates air pollution. The north has access to significant quantities of low-sulfur coal. The south does not, which will significantly increase the cost of controlling sulfur emissions.

---

[5]The Central China Power Grid is currently interconnected with the East China Power Grid, and additional interconnections with neighboring regional or independent grids are planned.

Finding 2. Regulating large emission sources requires very different policy instruments and has very different costs from regulating small emission sources.

Large sources with tall stacks contribute most to the long-range transport of sulfur emissions; small sources have even greater sulfur emissions, but they contribute mostly to local ambient concentrations in densely populated areas. Given the economies of scale—both technically and institutionally—regulatory regimes for large emission sources, such as power plants and key specialty industries, can greatly reduce total emissions, long-range transport, and impacts, depending on how close they are to major urban areas, as the case study in Changsha has shown. For small residential and commercial coal users, restricting or banning coal use in urban areas has proven to be an effective way of addressing ambient SO_2 pollution. Such measures have been most successful when they are part of a cross-sectoral plan that includes the widespread provision of cleaner fuels (e.g., natural gas) and the relocation of industry.

Finding 3. A better scientific understanding of the impacts of sulfur emissions and better estimates of the relative benefits of different control options will be very important for planning and implementing local control regimes.

The case studies indicate that local governments are usually able to identify sulfur-control measures to meet the emission-reduction targets of the central government. However, local governments need help in several areas: (1) determining whether control measures would achieve targets for ambient SO_2 concentration levels; (2) analysis of the impacts of sulfur pollution, including acid-rain hot spots; and (3) analysis of the cost effectiveness of control measures.

Finding 4. Promoting policies that have multiple benefits is an effective way of cutting sulfur pollution without relying on regulatory policies or institutions.

The marked decline of ambient SO_2 levels in many Chinese cities over the past five years is largely the result of reduced coal consumption among small-scale users made possible by widespread investments in natural gas and systematic relocations of industries. Thus, changing the fuel mix is part of a larger urban-development strategy that many local governments have embraced in the last decade or so.

LESSONS FOR NATIONAL POLICY

These case studies provide valuable information not only for local decision making, but also for the central government's decisions for the TCZ policy. The successes of China's sulfur-control policy include: (1) the introduction of restrictions on the production of high-sulfur coal; (2) requirements that coal-fired power plants switch to low-sulfur coals or install emission-control equipment; and (3) the setting of targets for reducing emissions. In anticipation of the implementation of pollution controls, SEPA has effectively mobilized local environmental agencies to begin planning and preparations. Local environment and municipal authorities have not only begun collecting sulfur-emission information, but they have also instituted restrictions on coal-burning devices and, in some localities, banned coal burning outright in densely populated areas.

Setting simple, clear goals at the national level and letting local governments and line agencies work out the details of implementation has been an effective strategy for sulfur control that reflects the administrative and bureaucratic structure of China's governmental system. The system works well when both the central and local governments are committed to achieving results, which appears to be the case for sulfur control.

RECOMMENDATIONS

Recommendation 1. SEPA should undertake a study of the long-term (20-year horizon) targets and goals for controlling sulfur pollution in China and assess the country's needs and efforts accordingly. SEPA should focus on understanding the dynamics of long-term sulfur emissions and the impacts of specific hot spots of sulfur emissions on ecosystems, agriculture, and human settlement areas. Regulatory policy should then be directed toward controlling emissions from key polluters and economic sectors and on avoiding the creation of hot spots.

Recommendation 2. SEPA should continue to provide scientific evidence of the impacts of sulfur, especially from thermal power plants, to reduce the interregional transport of emissions and thereby help the power sector comply with national sulfur regulations in the most economical way. The power sector is likely to determine the long-term success of China's sulfur pollution-control program because of its projected growth and the general softening of, or even reduction in, the demand for coal in other economic sectors. SEPA should also continue to monitor sulfur emissions in the transportation sector, a minor contributor in most Chinese cities.

Recommendation 3. Regulations for small emission sources should be kept simple and straightforward and should rely on cross-sector policy support to help eliminate clusters of small sources in urban areas. This will require the development of natural gas transmission lines and local investments in distribution facilities. The provision of gas for scattered coal-fired space-heating systems in northern cities is an important option. Large reductions in coal consumption for heating can also be made through a rapid scale-up in the development of energy-efficient buildings in northern China.

Recommendation 4. The focus on large emission sources and key industries should be continued. A permit system would reduce regulatory uncertainties and would probably reduce the costs of compliance. This will require a substantial increase in institutional and regulatory capacity at the provincial and municipal levels. A permit system is not simply a way of allocating emission quotas; it also includes a host of regulatory requirements on emissions and compliance, as well as consequences for violation. Such a system would pave the way for the introduction of a tradable permit system in the future.

Recommendation 5. The central government should provide assistance to localities to enable them to carry out the type of analysis done in Shijiazhuang and Changsha. Capacity building should be focused on the development of skills and institutions to: (1) assess and quantify the impacts of sulfur emissions; (2) evaluate the benefits of control options, including reductions in ambient concentrations of sulfur and associated impacts; and (3) assess the cost effectiveness of control options to determine which options will have multiple benefits.

REFERENCES

Akbar, S., and M. Kojima. 2003. Health Impacts of Outdoor Air Pollution. Urban Air Pollution, South Asia Urban Air Quality Management Briefing Note No. 11. Washington, D.C.: World Bank.

Downing, P., R. Ramankutty, and J.J. Shah. 1997. RAINS-ASIA: An Assessment Model for Acid Deposition in Asia. Washington, D.C.: World Bank.

ECON (Center for Economic Analysis). 2002. Review: An Environmental Cost Model. Oslo, Norway: Center for Economic Analysis.

EIA (Energy Information Administration). 2002. International Energy Outlook 2002. Washington, D.C.: U.S. Department of Energy.

Guttikunda, S.K., G.R. Carmichael, G. Calori, C. Eck, and J-H. Woo. 2003. The contribution of megacities to regional sulfur pollution in Asia. Atmospheric Environment 37(1): 11–22.

Hao, J. 2000. Dispersion Modeling and Damage Cost Valuation in China: A Case Study in Hunan Province. Beijing: Tsinghua University.

Li, J. 1998a. Benefit/Cost Analysis of Environmental Protection Input and Pollution Damage in Shijiazhuang. Shijiazhuang, China: Shijiazhuang Environmental Protection Bureau. In Chinese.

Li, J., J. Schwartz, and X. Xu. 1998b. Health Benefits of Air Pollution Control in Shenyang, China. Research Report. Cambridge, Mass.: Harvard University School of Public Health.

Li, J., S.K. Guttikunda, G.R. Carmichael, D.G. Streets, Y.S. Chang, and V. Fung. 2003. Quantifying the human health benefits of curbing air pollution in Shanghai. Journal of Environmental Management 70(1): 49–62.

NIVA (Norwegian Institute for Water Research). 2002. IMPACTS, Integrated Monitoring Program on Acidification of Chinese Terrestrial Systems. Oslo: Norwegian Institute for Water Research. Also available online at: *http://www.impacts.net.cn/*.

Oskarsson, K., A. Berglund, R. Deling, U. Snellman, O. Stenback, and J.J. Fritz. 1997. A Planner's Guide for Selecting Clean-Coal Technologies for Power Plants. Washington, D.C.: The World Bank.

Pu, Y., J.J. Shah, and D.G. Streets. 2000. China's "two-control-zone" policy for acid rain mitigation. Environment Management 24: 32–35.

SEPA (State Environmental Protection Administration). 2002. The 10th Five-year Plan for Prevention and Control of Acid Rain and SO_2 Pollution in the Two-Control Zones. Beijing: State Environmental Protection Administration.

SEPA. 2003. Plans and Reports. Available online at: *http://www.zhb.gov.cn/649364983179640832/index.shtml.*

Streets, D.G., T.C. Bond, G.R. Carmichael, S.D. Fernandes, Q. Fu, D. He, Z. Klimont, S.M. Nelson, N.Y. Tsai, M.Q. Wang, J.-H. Woo, and K.F. Yarber. 2003. An inventory of gaseous and primary aerosol emissions in Asia in the year 2000. Journal of Geophysical Research 108(D21): Article # 8809.

Wang, J., J. Yang, Z. Ma, and S. Benkovic. 2000. SO_2 Control in China's Power Industry, in SO_2 Emissions Trading Program, US Experience and China's Perspective. Beijing: China Environmental Science Press.

World Bank. 2001. China: Air, Land, and Water. Environmental Priorities for a New Millennium. Washington D.C.: World Bank. Executive Summary available online at: *http://www1.worldbank.org/publications/pdfs/14937execsumm.pdf.*

World Bank. 2003. China: Air Pollution and Acid Rain Control: The Case of Shijiazhuang and the Changsha Triangle Area. ESMAP Paper. Washington D.C.: World Bank.

Yang, L., I. Stulen, L.J. Dekok, and Y. Zheng. 2002. SO_2, NO_x and acid deposition problems in China: impact on agriculture. Phyton Special Issue on Global Change 42: 255–264.

Energy and Environmental Impacts of Chinese Rural Vehicles[1]

DANIEL SPERLING and ZHENHONG LIN
Institute of Transportation Studies
University of California, Davis

More than 3 million Chinese rural vehicles (CRVs) were produced in 2002, three times the number of conventional passenger cars. These small, simple, indigenous vehicles are widely used in small cities and rural areas but are virtually unknown outside China. CRVs provide huge benefits in terms of mobility and economic development, but they are also highly energy inefficient and polluting. CRVs now consume about one-fourth of the diesel fuel in China. Increasing government regulation (mostly for emissions and safety) is having profound effects on the industry, with uncertain implications for the sale and globalization of rural vehicle technology.

In 1994, the Chinese government designated the automotive industry a "pillar" of economic development. In the Tenth Five-Year Plan (2000–2005), the government established a goal of widespread car ownership, and since then, intense efforts have been made to engage the international automotive industry (Gallagher, 2003; NRC et al., 2003). As a result, passenger car output has been increasing rapidly, from 0.6 million in 2000 to 1.06 million in 2002 (China National Bureau of Statistics, 2004).

In striking contrast, and virtually ignored, is the even larger number of small

[1]This paper is based on a larger report by D. Sperling, Z. Lin, and P. Hamilton. 2004. Rural Vehicles in China: An Exploratory Analysis of Technology, Economics, Industrial Organization, Energy Use, Emissions, and Policy. Research Report UCD-ITS-RR-04-1. Davis, Calif.: Institute of Transportation Studies, University of California, Davis.

three-wheel (3-w) and four-wheel (4-w) vehicles manufactured by domestic Chinese companies for use in small cities and rural areas. With virtually no governmental financial support, the production of these CRVs first exceeded 3 million per year in 1999 and reached an estimated 20 to 22 million in 2001 (China Automotive Technology and Research Center, 2000; Chinese Government Website, 2003a). The implications of these vehicles are huge—in terms of safety, energy use, air pollution, noise pollution, greenhouse gas emissions, and rural development. The English language literature provides very little information about CRVs (indeed, there is no accepted English name for them), and even in Chinese, information is sparse.

HISTORY OF CHINESE RURAL VEHICLES

The CRV industry arose as a result of early efforts by the Communist government to boost rural development. The small enterprises that emerged and flourished were largely independent of national and provincial governments. Although they were at a disadvantage because they received little (if any) financial support from government at any level, they benefited from being subject to few regulations and little intervention. The sales history of the CRV industry is shown in Figure 1.

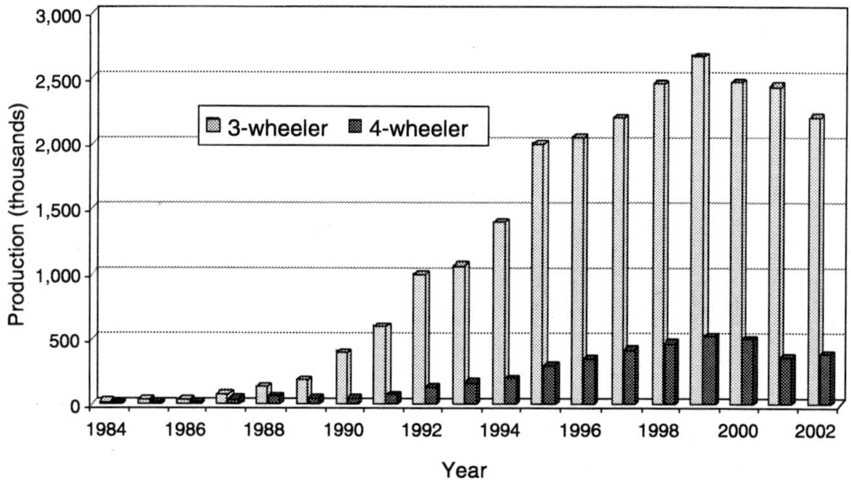

FIGURE 1 CRV production (in thousands), 1984–2002. Sources: Data through 1998 are from GM (1999); data for 1999–2000 are from Chinese Automotive Technology and Research Center (2000); data for 2001–2002 are from China Machinery Economic Information (2003).

As the CRV industry matures, it is also consolidating. In just one year, from 2001 to 2002, the number of registered CRV manufacturers dropped from 204 to 120, although it is believed that many manufacturers are still in business but no longer registered. In any case, the market share of the 10 largest manufacturers is increasing. In that same year, the top 10 3-w manufacturers increased their share from 59.5 percent to 65 percent, and the top 10 4-w manufacturers increased their share from 93 percent to 96 percent of the market. Two companies, Shifeng and Juli, accounted for 61 percent of the 3-w CRV market.

The government appears to favor consolidation in the CRV industry, as well as in the automotive industry, as a means of creating companies with greater resources and greater capabilities to develop and adopt advanced technologies, including emissions-control technologies (Harwit, 1995). The large number of companies in the CRV industry, ranging from small backyard shops to large industrial enterprises, is indicative of the traditionally low entry barriers in terms of capital investment and government licensing and rules. Price competition is severe, with strong downward pressure, and, until recently, companies had few incentives to invest in advanced technologies, especially for reduced emissions and other features that do not add much to consumer-perceived utility.

The new emission standards and policies being adopted by the central government will undoubtedly lead to further consolidations and the disappearance of small CRV manufacturers that produce poor quality vehicles and have no research and development (R&D) capabilities. In early 2003, Yanmar Co., Ltd., a Japanese company specializing in diesel engines, and Shandong Shifeng Group Co., Ltd., established a joint venture to produce and market single-cylinder diesel engines that comply with new CRV emission regulations. It remains to be seen how this collaboration will affect the health of the industry, the creation of a supplier industry, and industry responsiveness to the low end of the market.

The CRV industry is at a crossroads. As government regulation and intervention increase, as companies gain access to improved technologies (through internal R&D and transfer from others), and as the industry consolidates into fewer and larger companies, one would expect product quality to improve. Yanmar's penetration into the Chinese single-cylinder diesel-engine industry, the core of the 3-w CRV industry, is indicative of the changes in the configuration of the industry and the expectation of higher quality vehicles (Yanmar Co. Ltd., 2003).

DESCRIPTION OF CHINESE RURAL VEHICLES

CRVs are used mostly to transport goods. Powered by diesel engines, they are smaller and slower than conventional cars and trucks, and they use technology developed in China. Officially, the Chinese government treats the CRV industry as part of the farm machinery industry, rather than the automotive industry (defined by an official government standard, Technical Requirements on Safety for CRVs (GB18320-2001)). For a more detailed description that reconciles evolving technology with changing government rules and definitions, see Sperling et al. (2004).

CRVs range from simple 3-wheelers with one-cylinder diesel engines on a motorcycle-like frame that cost about $300 (Figure 2) to sophisticated, small, truck-like 4-wheelers that cost more than $5,000 (Figure 3). About 80 percent of the 22 million CRVs are powered by single-cylinder diesel engines originally designed for stationary agricultural machinery (China Agricultural Resources

FIGURE 2 Typical 3-w CRV. Source: Juli Company, 2003.

FIGURE 3 Sophisticated 4-w CRV. Source: Juli Company, 2003.

Network, 2003). These one-cylinder engines are very inefficient, especially in mobile applications, and emit large amounts of pollutants.

The preference for single-cylinder diesel engines is based on technology, economics, and policy. Diesel fuel is less prone to explode and burn than gasoline and is, therefore, safer. Diesel engines are also easier to maintain and can operate satisfactorily on poor quality fuels. With minimal mechanical skills, a farmer can repair a faulty injection pump, the principal source of diesel engine problems. In contrast, common maintenance problems with gasoline engines are related to carburetors and ignition systems, which require considerably more skill to repair.

Diesel engines are also easier to manufacture. In the 1970s and early 1980s, the only available small engines were single-cylinder diesel engines, which were widely used for walking tractors. Professor Liu (professor emeritus, Tsinghua University) has said, "Farmers themselves can even produce single-cylinder diesel engines. When CRVs became popular, many counties had their own single-cylinder diesel engine factories. One province could have tens of such factories."

Perhaps the most important factor in their popularity has been low fuel costs. Diesel engines use less fuel than gasoline engines because they are 20 to 30 percent more fuel efficient. In addition, the price of diesel fuel in rural China was one-third to one-half the price of gasoline until the 1980s. Since policies were changed in the mid-1980s, however, gasoline and diesel prices have been more comparable (CRSTA, 1982).

ENERGY USE

As a result of increased motorization, Chinese petroleum consumption has increased rapidly since 1990 (Figure 4). Increases in gasoline consumption reflect rising automobile use in urban areas and motorcycle use in urban and rural areas. Increases in the consumption of diesel fuel and other middle distillates are partly attributable to increasing diesel consumption in rural areas, including consumption by CRVs.

Unfortunately, it is difficult to determine the portion of energy use attributable to CRVs because reliable, detailed analyses are not available. Therefore, our estimates are based on fragmented information from the literature and our own knowledge of CRVs. We have reconciled bottom-up data based on the estimated number of vehicles and top-down data based on aggregate national estimates.

According to a study done by Xunying Yang (2001), total diesel fuel consumption in China in 2000 was 69.5 million tons (see Table 1), or 21.6 billion gallons. In his study, highway transportation accounted for 24 percent of total diesel fuel use, and CRV use accounted for 21 percent. Total gasoline consumption in China in 2000 was 35 million tons, or 12.5 billion gallons (CCCT, 2003).

Another study of oil consumption conducted by the China Petrochemical Consulting Corporation arrived at similar estimates for energy use by CRVs (although the assumptions and methods were not provided) (Ke and Shang, 2000).

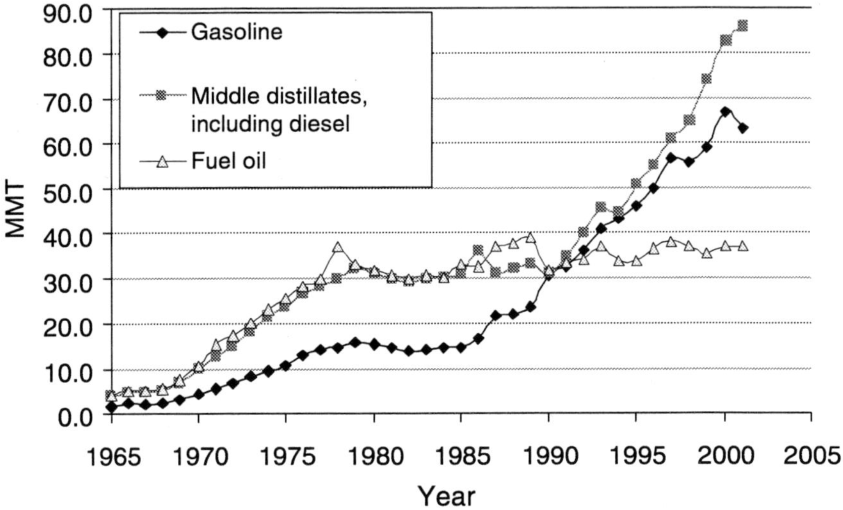

FIGURE 4 Chinese oil consumption in million metric tons (MMT), 1965–2004. Source: BP, 2003.

TABLE 1 Diesel Fuel Consumption in China (in millions of metric tons, MMT)

User	1996	1997	1998	1999	2000
Highway transportation	8.21	8.95	10.09	13.64	16.86
Light vehicle	1.16	1.57	2.02	1.93	1.83
Medium vehicle	5.55	5.60	5.95	8.18	10.10
Heavy vehicle	1.50	1.78	2.12	3.53	4.93
CRV	9.28	10.74	13.25	13.19	14.43
Railway	4.50	4.60	4.80	5.00	5.20
Marine	3.50	3.60	3.80	4.00	4.20
Agriculture	12.00	12.40	13.70	15.00	16.00
Fishing	5.70	6.00	6.10	6.30	6.50
Electricity	4.50	5.00	4.50	4.10	3.80
Others	2.50	2.50	2.50	2.50	2.50
Total	50.19	53.79	58.74	63.73	69.49

Source: Yang, 2001.

This study provided estimates for 1997 and forecasts for 2000 and 2005. The forecasts for the nonmilitary consumption of diesel fuel in 2000 was 26.6 million metric tons (MMT), somewhat lower than the 31.29 MMT estimated by Yang (Yang's estimate may include military transportation). In any case, the figures for diesel consumption by CRVs for 2000 were similar in the two studies: 14.28 MMT and 14.43 MMT.

For a bottom-up estimation of fuel use by CRVs, we relied on interviews with CRV users, fuel economy data provided by CRV makers (Sperling et al., 2004), expert opinion by Professor Liu, previous CRV regulations, and data on the numbers and composition of the CRV fleet. The parameters for the calculations for 2000 include: fuel economy of 2.8 liters per vehicle-ton per 100 km; average CRV weight (without payload) of 1.0 ton; average payloads of 0.5–1 ton; useful lives of six years for 3-w and single-cylinder 4-w CRVs and nine years for multicylinder 4-w CRVs; accumulated kilometers of travel of 120,000 km for 3-w and single-cylinder 4-w CRVs and 250,000 km for multicylinder 4-w CRVs; and a total vehicle fleet of 19 million CRVs.

Our estimate of diesel consumption by CRVs in 2000 is 19.03 MMT, if vehicles weigh 2 tons (including payload), or 14.27 MMT if they weigh 1.5 tons. The latter figure is close to those of the two studies cited above. Based on our analysis, the two aggregate studies based their estimates on company fuel economy data and the 0.5 ton payload. But based on the high frequency of overloading, we believe that 19.03 MMT is a more reasonable estimate. This would represent about 25 percent of total diesel consumption in China in 2000, equivalent to almost 50 percent of all of the gasoline consumed in the country.

EMISSIONS

Even less is known about CRV emissions than about CRV fuel consumption. Clearly, though, CRVs are high emitters of smoke and pollution. Even when new, these primitive single-cylinder diesel engines emit clouds of black smoke, particularly under heavy load and at low RPM, conditions frequently encountered because overloading is common and the vehicles have three-speed gearboxes.

No data are available on the contribution of CRVs to total pollution, but rough calculations and assumptions suggest that the total amount may be similar to the amount from conventional vehicles. This conclusion is based on the following information and estimates. First, one government document asserts that "CRVs in China are powered by diesels and their emissions per unit of energy are on average twice that of trucks" (Ji, 2003). The situation may be even worse for 3-w CRVs. Second, we combined that assertion with statistics indicating that 88 percent of the 2.8 million CRVs produced in 2001 were single-cylinder CRVs and that more than 60 percent of single-cylinder 3-w CRVs and about 30 percent of multicylinder 4-w CRVs cannot meet the emissions requirements of GB 18322-2002, Limits and Measurement Methods for Smoke at Free Acceleration from CRVs (Ji, 2003). Based on these assumptions, we estimate that the contribution of CRVs to air pollution in China is equivalent to that of all other motor vehicles combined.

Although CRVs are more widely dispersed than cars, in small towns where

CRVs are allowed and agricultural trade is active, CRV emissions are likely to have a large impact on local air quality. This pollution is damaging not only to human health, but also to agricultural production. Recent studies have suggested that air pollution in China in the form of ground-level ozone (Aunan et al., 2000) and atmospheric aerosols (Chameides et al., 1999) can substantially reduce crop yields. Nitrogen oxides (NO_x), a principal pollutant from diesel engines, is a major source of ozone; and small airborne particles, also from diesel engines, are a major source of aerosols, which can absorb sunlight and contribute to regional haze.

The Chinese government is beginning to pursue more aggressive policies to reduce emissions from motor vehicles. In 2002, the government announced that by 2004 diesel trucks must meet standards equivalent to Euro 2 standards (which took effect in the European Union in 1996) (China Environment News, 2003) and that CRVs must meet the equivalent of Euro 1 standards by 2005 (effective in the European Union in 1993) (Xu, 2003). At the same time, the government adopted rules and tests for CRV emissions, known as GB 18322-2002, Limits and Measurement Methods for Smoke at Free Acceleration from CRVs (Chinese Government Website, 2003b).

There is some uncertainty about the enforcement of these rules. On one large CRV maker's website, the only emission attribute specified for 3-w and 4-w CRVs is for visible smoke (Juli Company, 2003). No mention is made of other emissions, such as hydrocarbons, NO_x, carbon monoxide, and invisible particulate matter. From this, we concluded that only visible smoke is being regulated for CRVs.

The national emission law, GB 18322-2002, delegates some authority to provincial governments: "Smoke limits for CRV driving in developed urban areas can be determined by provincial governments." We are not certain about the interpretation of this provision, but apparently it is the legal basis for local governments to adopt their own policies regarding CRV use. Indeed, in many cities, diesel vehicles of all types are allowed only during the night (e.g., 9 p.m. to 6 a.m.); in others, CRV owners may purchase special license plates that allow them to enter the city. In cities where CRVs are allowed, the licenses and registrations are controlled and can be very expensive.

Even though emissions rules are lax for the CRV industry, they still present a substantial challenge for CRV manufacturers. Although we do not know how this will play out, the standards certainly put pressure on CRV companies to consolidate to support R&D on engines and emissions, as well as to seek investment and expertise from international carmakers and parts suppliers. Indeed, as the recent joint venture between the largest CRV company (Shifeng) and an international manufacturer of diesel engines (Yanmar) indicates, that process has already begun.

CONCLUSIONS

Developing countries rarely have the capabilities to design and mass produce functional vehicles at low cost without significant foreign assistance and public subsidy. Yet that is precisely what happened in China. The good news is that CRVs have been a boon to rural development in China. The downside is their high energy use and air pollution.

Chinese governments are now intervening to improve CRV technology. But these interventions are seen by the CRV industry as a mixed blessing. On the one hand, stricter requirements for safety, emissions, and vehicle quality will no doubt lead to better vehicles and will accelerate the weeding out of undercapitalized companies with weak technical capabilities. The companies that survive will be better able to compete in the world market. On the other hand, it remains to be seen whether or not the industry will serve the low-cost rural markets as well as it has and whether or not competing governmental interests will create an even more uncertain and risky financial and market environment for CRV makers.

In any case, the future of the CRV industry will play an important role, not only in China, but also potentially worldwide. CRVs have been central to rural development in China and could play an equally positive role in other developing countries. But the CRVs being built today are inefficient users of petroleum and large emitters of pollution and greenhouse gases.

Therefore, the future of CRVs is largely in the hands of the Chinese government. Will the government clamp down on the negative attributes of CRVs with more stringent rules and more aggressive enforcement? Or will it create incentives for their proliferation? The government must consider how to balance the benefits of rural development with the benefits of safety, energy, and environmental standards. The government must find a way to develop effective industrial policy to mitigate the negatives and promote the benefits of CRVs. The debate would be greatly helped by more research on costs and benefits and on the effectiveness of various industrial policy strategies and policy instruments.

ACKNOWLEDGMENTS

We thank Hanming Huang, an analyst at the Chinese Economic Information Network, and emeritus Professor Zheng Liu of Tsinghua University for providing us with valuable data and information. Professor Scott Rozelle of the University of California, Davis, provided valuable early assistance, and Feng An, Dennis Scheutzle, and Michael Quanlu Wang provided helpful suggestions. Seed funding was provided by General Motors, Toyota, DaimlerChrysler, and the Energy Foundation.

REFERENCES

Aunan, K., T.K. Berntsen, and H.M. Seip. 2000. Surface ozone in China and its possible impact on agricultural crop yields. Ambio 29(6): 294–301.

BP. 2003. Statistical Review of World Energy 2003. Available online at: *http://www.bp.com/subsection.do?categoryId=95&contentId=2006480*, accessed on 5/10/04.

CCCT (China Clean Coal Technology). 2003. Energy Consumption by Industry in 2000. Available online at: *http://www.cct.org.cn/cct/content.ASP?ID=5831*, accessed on 5/9/04.

Chameides, W.L., H. Yu, S.C. Liu, M. Bergin, X. Zhou, L. Mearns, G. Wang, C.S. Kiang, R.D. Saylor, C. Luo, Y. Huang, A. Steiner, and F. Giorgi. 1999. Case study of the effects of atmospheric aerosols and regional haze on agriculture: an opportunity to enhance crop yields in China through emission controls? Proceedings of the National Academy of Sciences 96(24): 13626–13633.

China Agricultural Resources Network. 2003. The Small Diesel Engine Industry Needs Technology Breakthrough. Available online at: *http://sdnj.gov.cn/data/2002/tg/1115214.htm*, accessed on 6/17/03. In Chinese.

China Automotive Technology and Research Center. 2000. The 2000 Chinese Automotive Industry Yearbook. Beijing: China Machine Press. In Chinese.

China Environment News. 2003. Chinese Diesel Motor Vehicles Heading for Euro 2. Available online at: *http://www.envir.gov.cn/info/2003/3/325119.htm*, accessed on 7/12/03. In Chinese.

China Machinery Economic Information. 2003. The Output of Chinese Rural Vehicles Keeps Decreasing in 2002. Available online at: *http://www.meicon.com.cn/*, accessed on 3/26/03.

China National Bureau of Statistics. 2004. Yearly Statistics Data. Available online at: *http://www.stats.gov.cn.*

Chinese Government Website. 2003a. Notification about the Approval by PRC State Council on the Proposal of Abolishing Allowance Policy for Rural Diesel Consumption. Available online at: *http://www.china-njx.com/zcfg/040.htm*, accessed on 7/23/03. In Chinese.

Chinese Government Website. 2003b. Limits and Measurement Methods for Smoke at Free Acceleration from Agricultural Vehicles. State Environmental Protection Administration of China. Available online at: *http://www.sepa.gov.cn/6502083043031777728/20030310/1037159.shtml,* accessed on 5/7/03. In Chinese.

CRSTA (China Rural Special Technology Association). 1982. Notification about the Approval by PRC State Council on the Proposal of Abolishing Allowance Policy for Rural Diesel Consumption. Available online at: *http://www.china-njx.com/zcfg/040.htm*, accessed on 7/23/03. In Chinese.

Gallagher, K.S. 2003. Foreign Technology in China's Automobile Industry: Implications for Energy, Economic Development, and Environment. China Environment Series no. 6. Washington, D.C.: Woodrow Wilson Center.

Harwit, E. 1995. China's Automobile Industry: Policies, Problems, and Prospects. Armonk, N.Y.: M.E. Sharpe.

Ji, G. 2003. The Serious CRV Pollution Issue Pushing the Government to Effort in Regulation. Available online at: *http://www.sepa.gov.cn/6490966894575616000/20020702/1011810.shtml,* accessed on 4/7/03. In Chinese.

Juli Company. 2003. Product Introduction Webpage. Available online at: *http://www.julisuliao.com,* accessed on 7/11/03. In Chinese.

Ke, X., and C. Shang. 2000. Development Trend of Domestic Civil Vehicles and Its Influences on Supply of Oil Products. Beijing: China Petrochemical Consulting Corporation. Also available online at: *http://www.spc.com.cn/spcspc/Chinese/tep/2000_04/7.html*, accessed on 6/29/03. In Chinese.

NRC (National Research Council), NAE (National Academy of Engineering), and Chinese Academy of Engineering. 2003. Personal Cars and China. Washington, D.C.: The National Academies Press.

Sperling, D., Z. Lin, and P. Hamilton. 2004. Rural Vehicles in China: An Exploratory Analysis of Technology, Economics, Industrial Organization, Energy Use, Emissions, and Policy. Research Report UCD-ITS-RR-04-01. Davis, Calif.: Institute of Transportation Studies, University of California, Davis.

Xu, H. 2003. Market Can't Compete with Environmental Protection. Available online at: *http://www.qiche.com.cn/files/200204/02048.html*, accessed on 6/26/03. In Chinese.

Yang, X. 2001. Analysis and forecast of China's gasoline, kerosene and diesel fuel market. Petroleum and Petrochemicals Today 9(11): 13–18. In Chinese.

Yanmar Co. Ltd. 2003. Yanmar Agrees to Establish a Joint Venture with Shandong Shifeng Group, China's Largest Farm Vehicle Manufacturer. Available online at: *http://www.yanmar.co.jp/english/aboutus/whats-new/news/0301/conts01.htm,* accessed on 6/19/03.

Global Impacts

Atmospheric Long-Range Transport of Urban Pollutants

LEONARD LEVIN
Electric Power Research Institute

Atmospheric processes, including the wind-borne transport of air pollutants, are typically classified into time and equivalent spatial scales, each of which includes dominant physical and chemical processes. For example, the "mesoscale-beta" spatial scale typically extends from 200 to 2,000 km in space and from six hours to two days in time. Dominant processes at this scale are land-sea differences in heating, large topographic features, and atmospheric centers of action (or high and low pressure systems and the frontal patterns between them). The equivalency of time and space imply mean wind speeds of about 30 km/h for transport of material, representing frontal pattern transport rather than geostrophic motion of centers of action.

Long-range transport of pollutants is generally taken to mean transport over spatial scales in the mesoscale-beta range or larger. During this transport, the pollutant(s) of primary interest, as well as the co-pollutants transported with them, may undergo chemical and photochemical transformation. Methods of relating pollutant levels measured at distant receptor points to their area of origin and transport route vary widely. One method uses co-pollutant tracers; for example, elevated sulfate levels may indicate an industrial source zone and "tag" the location of origin for co-transported mercury. Selenium is used as an indicator for oil or coal combustion, and elevated levels of selenium are commonly used to "tag" chromium or nickel.

Global industrial development has resulted in concentrations of energy, production, and commercial facilities in urban areas and consequent concentrations of emissions of complex suites of organic and inorganic atmospheric pollutants into general atmospheric circulation. This paper is an overview of the teleconnections between densely populated areas via the transport of emitted pollutants. Urban areas are (1) significant source areas for pollutants in distant receptor points and (2) receptor areas for pollutants originating at great distances that often have a significant effect on the baseline air quality in the receptor areas.

LONG-RANGE TRANSPORT AND URBAN AREAS IN CHINA

Over the last decade, a number of international bodies, such as the International Institute for Applied Systems Analysis (IIASA) and the United Nations Environment Program (UNEP), have accumulated a body of data and model results demonstrating the significant contributions of combustion and mobile sources in China to air quality at distant points. Data collected throughout the 1990s and the first part of this century have shown the predominance of urban sources for a wide range of pollutants. For example, during the multinational ACE (Asian Pacific Regional Aerosol Characterization Experiment)-Asia experiment in 2001, Clarke et al. found that aerosol measurements over the Yellow Sea indicated the presence of large urban areas upwind; the highest values of both light scattering and absorption occurred below 1 km elevation, becoming moderate scattering but high absorption between 1 and 4 km, and dropping off rapidly above 4 km. This pattern of light modification is characteristic of high concentrations of both primary (large) and secondary (smaller, $< 0.75\mu m$) aerosols in the lowest layer, a signature of urban emissions from combustion and vehicular use; the middle layer shows chiefly larger primary particles, potentially representing elevated emissions from combustion sources.

Numerous incidents have been reported of Asian contributions to particulate matter or ozone on the western boundary of North America, some of them attributable to increased emissions from or greater production downwind of urban areas in Asia (Holloway et al., 2003). Recent data are compiled in Tables 1 and 2.

Asia, particularly China, is more problematic as a receptor area for emissions from other continents, partly because of the quantity of emissions just upwind of the Asian landmass, partly because of the monsoonal circulation created by the size of that landmass, and partly because of the large-scale general circulation of the global atmosphere.

Generally, China is in the downwind direction of sources in the prevailing mid-latitude westerlies, but somewhat south of the path of centers of action that redistribute either persistent pollutants or secondary products. South China lies astride the "horse latitudes" marked by generally easterly flow, specifically the western extension of the El Nino-Southern Oscillation (ENSO) pattern of the cisequatorial Pacific basin. Monsoonal patterns over east Asia contribute to the strong transport of emissions from industrialized Europe to the northeast. Lester Machta of the U.S. National Oceanic and Atmospheric Administration noted high levels of sulfates in the (then) Soviet arctic north of the Yenisei River delta and

TABLE 1 Surface Aerosol Increases at Northern Mid-latitudes from Intercontinental Transport of Pollution

Source Region	Receptor Region	Aerosol Type	Aerosol Enhancement above Long-Term Baseline ($\mu g/m^{-3}$ unless stated otherwise)	Method of Estimate
Asia (mean)	U.S., yearly means	organic carbon	0.013 (western U.S.) 0.007 (eastern U.S)	Sensitivity simulation with no anthropogenic emissions from source region.
Elemental carbon	0.005 (western U.S.) 0.003 (eastern U.S.)			
Asia (events)	Northwestern U.S., spring 1997	all	~200 particles cm^{-3}	Observed enhancements at Cheeka Peak Observatory in air masses of Asian origin.
Asia (dust event)	Western U.S., April 1998	all	40–63 (PM_{10}), 4–11 ($PM_{2.5}$)	Observed enhancements at a number of monitoring stations.
Asia (dust event)	Lower Fraser Valley, British Columbia, Canada, April 1998	all	18–26 (PM_{10})	Attribution based upon elemental composition at a number of monitoring stations.
Asia (dust event)	Northwestern U.S., April 1993	all	4–9 (PM_{10})	Observed enhancements at three monitoring stations.
Asia	Western U.S., spring, summer, fall 1989–1999	all	0.2–1 ($PM_{2.5}$), rare exceedances of 5	Attribution based upon matching Asian source type (diagnosed from April 1998 events) via cluster analysis at a number of monitoring stations.

Source: Holloway et al., 2003, and references therein.

TABLE 2 Surface Ozone Enhancements at Northern Mid-latitudes from Intercontinental Transport of Pollution

Source Region	Receptor Region	O_3 Enhancement (ppbv)	Method of Estimate
Asia	Northwestern U.S., spring	4 (mean), 7.5 (max)	Sensitivity simulation with no anthropogenic emissions from source region.
Asia	Western U.S., spring	3–10 (range during Asian pollution events)	Sensitivity simulation with no surface emissions from source region.
Asia	Europe, U.S.	1.0 (U.S.), 0.8 (Europe)	Annual mean enhancements from sensitivity simulations with 10 percent increases in emissions from source region. Results were multiplied by 10 to estimate total effect of current anthropogenic emissions from the source continent.
Europe	Asia, U.S.	1.1 (Asia), 0.9 (U.S.)	
U.S.	Europe, Asia	2.0 (Europe), 0.8 (Asia)	
Asia and Europe	U.S., summer	4–7 (typical afternoon range), 14 (max)	Sensitivity simulation with no anthropogenic NO_x and non-methane VOCs emissions from source region.
Europe	East Asia, spring	3 (daytime mean)	Ibid.
North America	Europe, summer	2–4 (daytime mean), 5–10 (events)	Ibid.
Europe	East Siberia	2 (annual) 3 (spring-summer)	Difference between median observed O_3 concentrations in 1997–1999 for air masses originating in Europe vs. Siberia and high latitudes.

back-trajectory analysis to their sources in Central Europe (personal communication). As noted by Martin (2003):

> Modelling studies suggest that the relative contribution of emissions sources on other continents to annual mercury deposition is largest for North America, less for Europe, and lowest for Asia, which is primarily due to the spatial pattern of emissions.

TABLE 2 Continued

Source Region	Receptor Region	O_3 Enhancement (ppbv)	Method of Estimate
North America	Mace Head, Ireland	0.4 (winter) 0.2 (spring) –0.3 (summer) –0.9 (fall)	Mean observed difference in O_3 concentrations in 1990–1994 for air masses originating in the U.S. and Canada vs. Iceland and Greenland.
North America	Europe, yearly mean	18 (Atlantic fringes) 10–15 (central Europe)	Ozone produced in tropospheric column over source region.
Asia	Europe, yearly mean	9 (Atlantic fringes) 5–7 (central Europe)	
Background (anthropogenic methane)	U.S., summer	6 (afternoon mean)	Sensitivity simulation with anthropogenic CH_4 emissions reduced globally by 50%. O_3 enhancements from that simulation were doubled to estimate total enhancement from anthropogenic CH_4.
Background (1980–1998)	U.S.	3–5 (spring, fall)	Observed trend in the lower quantiles of the O_3 frequency distribution at rural sites.
Background (1984–2002)	U.S. west coast	10	Observed trend at surface sites and from aircraft missions (1984–2002).
Asia (future)	U.S.	2–6 (western U.S.) 1–3 (eastern U.S.) highest in April–June	Sensitivity simulation with tripled Asian NO_x and non-methane VOCs emissions.
Asia (future)	Western U.S., spring	30–40 (max during Asian pollution events)	Sensitivity simulation with quadrupled Asian emissions.

Source: Holloway et al., 2003, and references therein.

MERCURY AS A TRACER SUBSTANCE

The development of sensitive, rapid-response instrumentation in the last five years has enabled us to consider using measurements of mercury as tracers for source regions and, in some cases, for source points. The widespread use of the Tekran™ device allows ground and aircraft measurements of total gaseous mercury (TGM) and speciated mercury at ambient levels in the environment. The Tekran instruments work by pumping ambient air with contained mercury vapor into the analysis module. During part of each sampling cycle, at lower temperature, the reactive gaseous mercury (RGM) is captured on an annular denuder while the remaining elemental mercury passes through the analyzer; during a desorption phase at higher temperature (500ºC), the analyzer alone pumps at a lower rate while the RGM is driven off the denuder and reduced, allowing measurement of the RGM portion of TGM.

Because of the relatively slow reaction rate of mercury in the atmosphere, mercury measurements can be used to track occurrences of higher mercury concentrations above background, both downwind to receptor regions and upwind to source regions. Mercury can thus be considered a conservative tracer substance, at least in the atmosphere.

Because of its complex biogeophysical cycling, however, there is as yet no good way to use mercury as a tracer from an atmospheric source to its ultimate uptake by food fish in the upper trophic level. There is some evidence that naturally occurring, lighter, stable isotopes of mercury may become slightly enriched through natural cycling due to the high vapor pressure of elemental mercury in aqueous environments (EPRI, 1998). William Landing of Florida State University is currently working to compare natural isotope ratios of mercury in coal at operating power plants with ratios in nearby waterways.

The Asian continent as a whole, and China in particular, is currently the single largest source of mercury emissions to the atmosphere from human activity (Pacyna et al., 2002). Table 3 shows this dominance. It is estimated that China contributes 500 Mg (1 Mg = 10^6 g = 1 metric ton [MT]) of total mercury from small, distributed industrial furnaces and village-level coal burning alone, or about 25 percent of the global emissions from anthropogenic sources.

During 2001, components of the ACE-Asia experiment were carried out jointly by agencies of the United States, Japan, Canada, South Korea, and China. One component of this extensive study was a mercury-tracking experiment by Electric Power Research Institute (EPRI), carried out by scientists of the National Center for Atmospheric Research (NCAR), in Boulder, Colorado. In this experiment, an NCAR C-130 aircraft equipped for extended high-altitude flight and using a Tekran 1130 analyzer sampled concentrations of TGM while flying transects off the east coasts of Japan and south China. During these flights, cotracer substances characteristic of presumed source categories were also measured.

Figure 1 shows the suite of research flights undertaken in the NCAR C-130 aircraft. Figures 2 and 3 illustrate a transect of the emission plume from the Miyake Jima volcano, a natural background source on the east coast of Honshu Island, Japan. Joint measurement of sulfur dioxide (SO_2) using a fast-response SO_2 analyzer illustrates the center of the plume. The "square wave" pattern of mercury concentrations is due to the integration time of the Tekran analyzer.

Figure 4 illustrates the signature pattern of a mix of urban air emissions from the industrialized area of southeast China near the city of Shanghai. These

TABLE 3 Global Emissions of Mercury from Anthropogenic Sources for 1995 (in metric tons = 10^6 g)

Continent	Stationary Combustion	Non-ferrous Metal Production	Pig Iron and Steel Production	Cement Production	Waste Disposal	Total
Europe	185.5	15.4	10.2	26.2	12.4	249.7
Africa	197.0	7.9	0.5	5.2		210.6
Asia	860.4	87.4	12.1	81.8	32.6	1,074.3
North America	104.8	25.1	4.6	12.9	66.1	213.5
South America	26.9	25.4	1.4	5.5		59.2
Australia/Oceania	99.9	4.4	0.3	0.8	0.1	105.5
TOTAL[a]	1,474.5	165.6	29.1	132.4	111.2	1,912.8

[a]In addition, emission of about 514 metric tons of mercury was estimated from chlor-alkali plants, gold production, and mercury used for various purposes (e.g., primary battery production, production of measuring and control instruments, production of electrical lighting, wiring devices, and electrical switches, etc.) in 1995. Source: Pacyna et al., 2002.

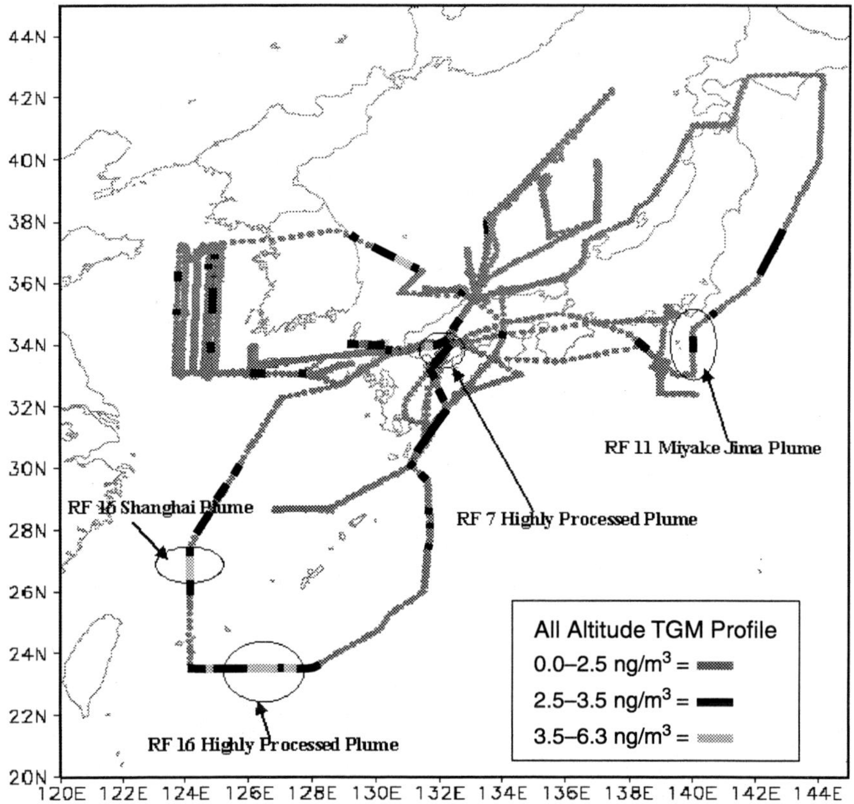

FIGURE 1 GIS-located flight tracks for the 2001 ACE-Asia Experiment mercury runs by NCAR C-130 aircraft. Circled segments indicate the plume of TGM from Shanghai, China, "aged" by residence in the marine boundary layer during over-water fetch. TGM concentrations in circled segments range from 3.5 to 6.3 ng/m^3 TGM (integrated over air column).

indications of source strength and characteristics were extended out into the Pacific basin. Figure 5 is a depiction of the three-dimensional structure of the research flights east of the south China coast, which were able to track the Shanghai urban plume for some 600 km over the Pacific.

As significant as these findings were, the receipt of the plume material on the west coast of North America remained to be investigated. In May 2003, flights were undertaken from Monterey, California, on a transect offshore California, Oregon, and Washington state (Friedli et al., 2004). Figure 6 illustrates simultaneous measurements of altitude above sea level, TGM in ng/m^3, and carbon

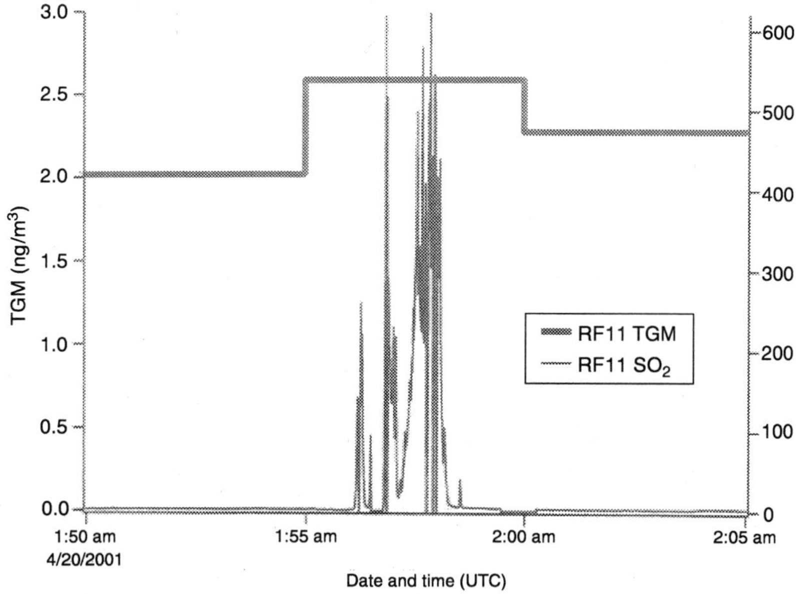

FIGURE 2 Instrument traces for TGM and SO_2 during transect of downwind plume from Miyake Jima volcano, Honshu Island, Japan, April 20, 2001. Response time for SO_2 sensor (95 percentile rise time) was typically 10 ms. Integration time for Tekran 1130 gaseous mercury sample was five minutes.

monoxide (CO, a tracer for urban area plumes) in parts per billion volume (ppbv). The Shanghai mercury plume is evident in the marine boundary level, even at these distances, some 1.6×10^4 km from the source.

An even higher concentration of mercury is evident at some 8,000 feet of elevation. Figures 7 and 8 show back-trajectory calculations indicating the originating region of the lower plume as the south China Sea; the upper plume is traced as far as central Asia. Additional data for this higher-elevation plume might have allowed tracking of the plume origins further west on the Eurasian land mass, perhaps to source regions in eastern Europe (there are no known high emission rate mercury sources in central Asia).

CONCLUSIONS

In recent years, researchers have noted numerous instances of both transatlantic and transpacific transport of marker pollutants. Measurements have been made of crustal material from Central Asia and the Sahara, ozone resulting from the photochemical transformation of precursor material from Asian urban regions,

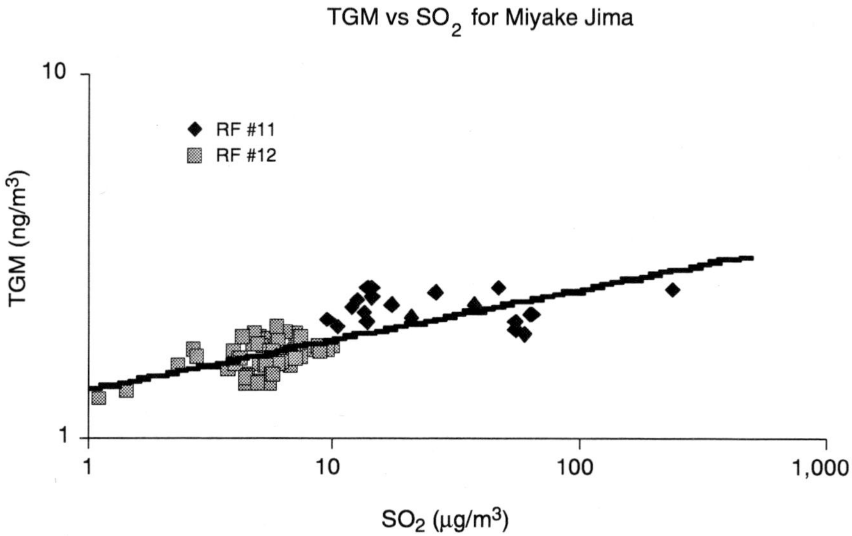

FIGURE 3 Correlation between SO_2 in µg/m^3 and TGM in ng/m^3 for the Miyake Jima volcano, demonstrating the close correlation between the two for this set of two research flights and the potential for tracking mercury from this category of background source in isolated locations.

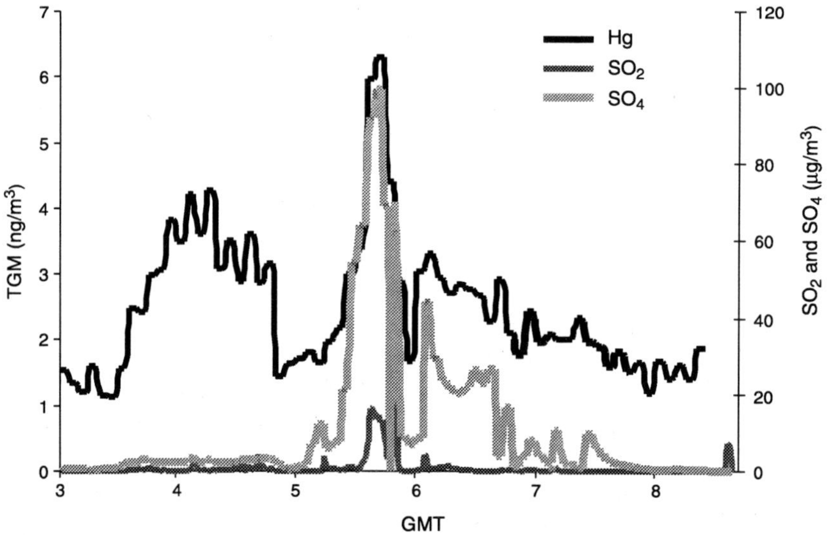

FIGURE 4 Instrument traces for SO_2, SO_4, and TGM downwind to Shanghai, China. Mercury (topmost trace) scale is on left vertical axis; sulfur species are on right axis. Note the difference in time scale from Figure 2.

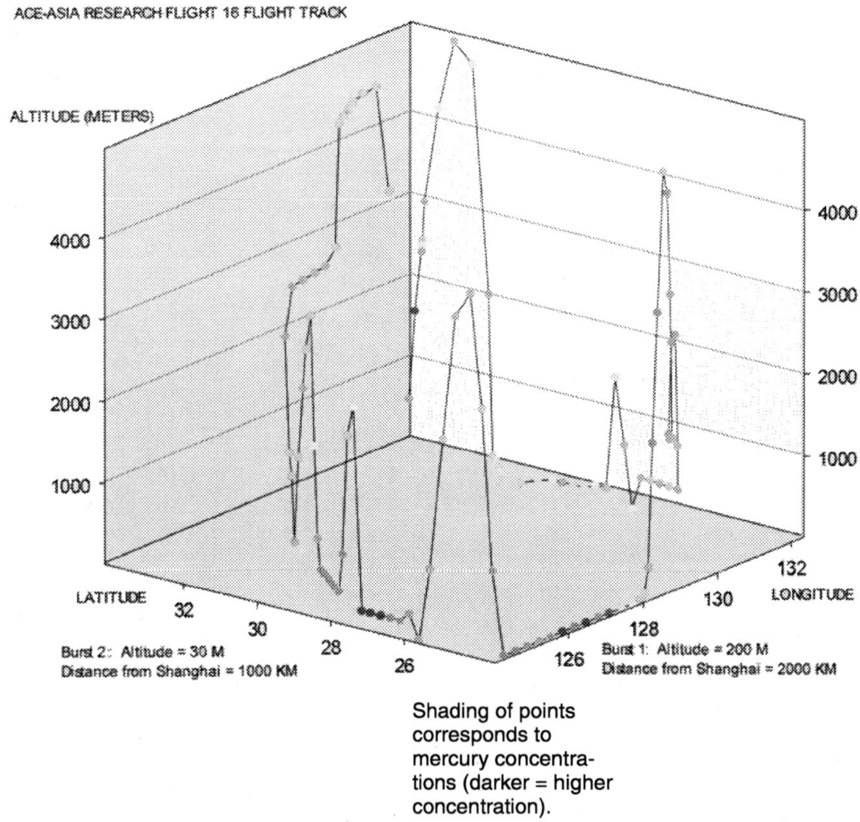

FIGURE 5 Demonstration of Shanghai plume "aging" in marine boundary layer. Three-dimensional depiction of flight track over the southwestern North Pacific tracking a mercury plume from Shanghai, China. Colors of data points represent concentrations of TGM.

and particulate matter with associated urban marker substances across the tropical Atlantic. With recent advances in measurement methods and sample analysis, mercury can be used as a tracer for source areas far upwind, in some cases across the Pacific.

Measurements by aircraft in 2001 and 2002 were able to track mercury from emission regions in industrialized southeast China, near Shanghai, and later detect the same mix of mercury and co-pollutants penetrating the west coast of the United States. Based on co-emitted characteristic co-pollutants, these measurements were able to discriminate between urban anthropogenic sources of mercury and background natural sources. As the discrimination of stable mercury isotope ratios improves, these methods may be combined to allow subcategorization of

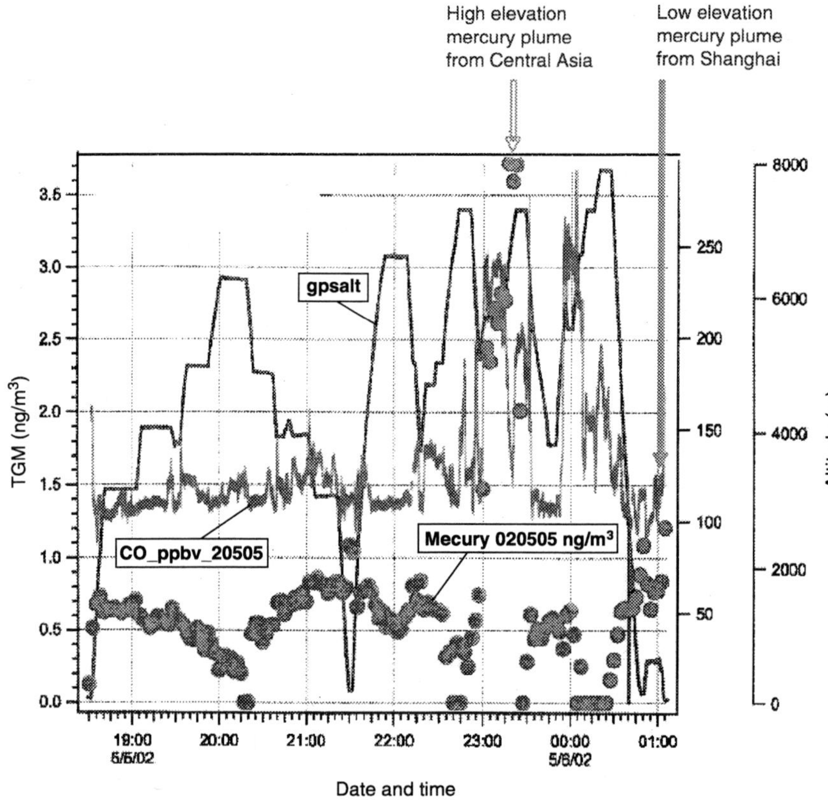

FIGURE 6 Evidence of dual mercury plumes from Asian landmass penetrating west coast of United States, May 5–6, 2002 (hours zulu), transect off San Diego, California, north to mouth of the Columbia River, Oregon/Washington. Mercury in ng/m^3 on left scale, CO in ppbv on inner right scale, altitude MSL in meters on outer right scale. Source: Friedli et al., 2004.

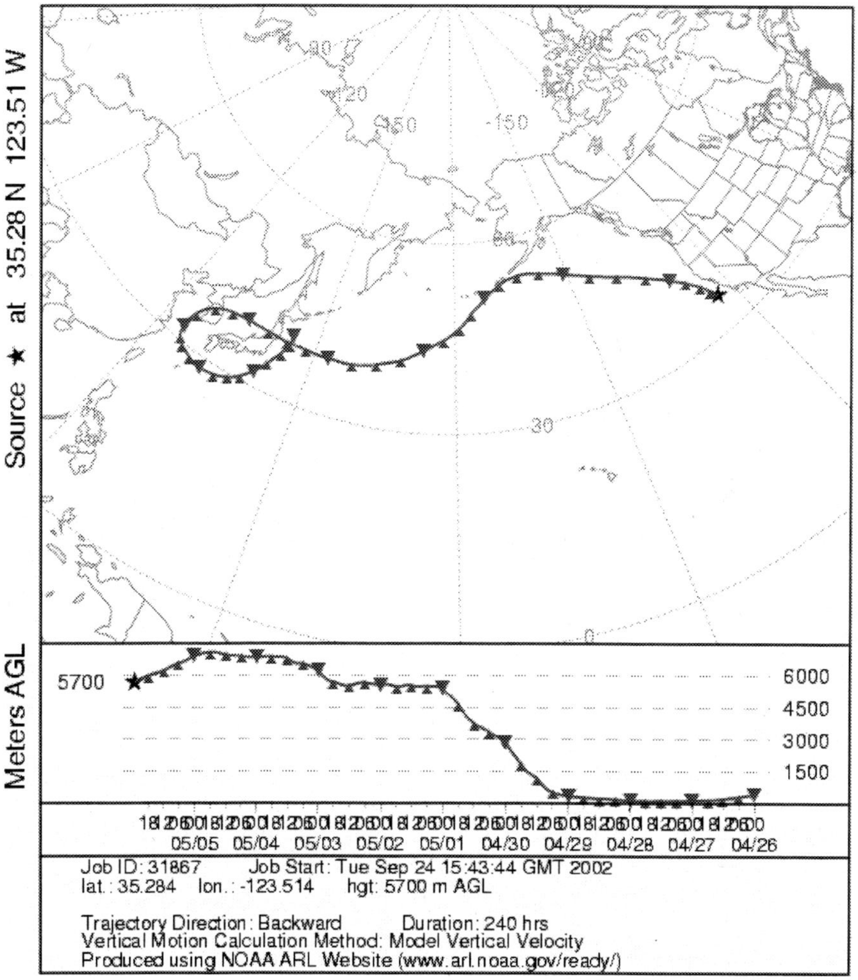

FIGURE 7 Back-trajectory calculations by U.S. National Oceanic and Atmospheric Administration for lower urban mercury plume, 10-day traverse of North Pacific from South China Sea/Sea of Japan origin region (backward trajectory ending at 23 UTC 05 May 02, FNL meteorological data). Note lofting of plume centerline at approximately day four due to intrusion of colder marine air mass in central North Pacific.

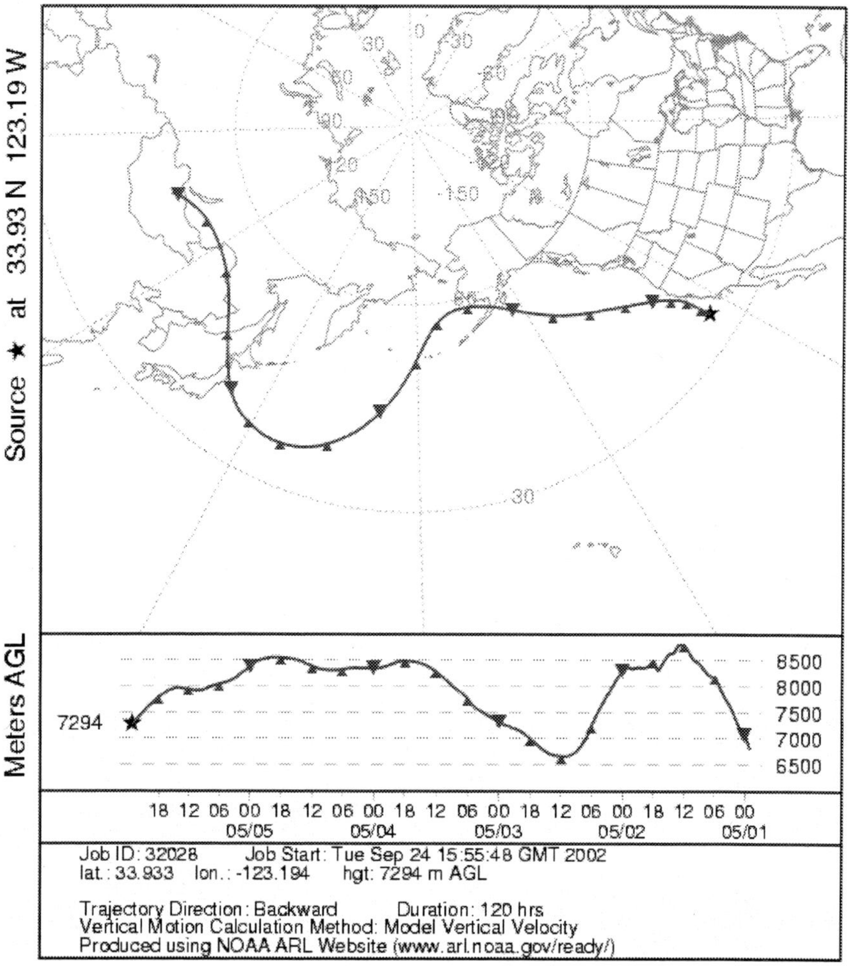

FIGURE 8 Back-trajectory calculations by the National Oceanic and Atmospheric Administration for upper Central Asia mercury plume, five-day traverse of eastern China and North Pacific from Mongolian plateau origin region (backward trajectory ending at 23 UTC 05 May 02, FNL meteorological data). Data limitations on wind structure in west-central Asia are presumed to limit further distinction of plume traverse from potential source regions further west in Eurasia. Note change in vertical scale on lower left altitude trace; the plume origin height is consistent with a source in central Asia (within data limits). Drop in elevation at approximately day two is consistent with entrainment into cyclonic area in western Pacific. Source: Friedli et al., 2004.

sources from distant measurement locations and improved source-receptor analyses for suites of pollutants.

REFERENCES

Clarke, A.D., Y. Shinozuka, V.N. Kapustin, S. Howell, B. Huebert, S. Masonis, T. Anderson, D. Covert, R. Weber, J. Anderson, H. Zind, K.G. Moore II, and C. McNaughton. 2001. Size-Distributions and Mixtures of Black Carbon and Dust Aerosol in Asian Outflow: Physiochemistry, Optical Properties, and Implications for CCN. Available online at: *http://saga.pmel.noaa.gov/aceasia/pubs/ClarkeLightAbsTrAATxtMar03.pdf* (accessed April 15, 2004).

EPRI (Electric Power Research Institute). 1998. The Use of Stable Mercury Isotopes and Surrogate Tracers to Investigate Source-Receptor Relationships. Report TR-111977. Palo Alto, Calif.: EPRI.

Friedli, H.R., L.F. Radke, R. Prescott, P. Li, J. Woo, and G.R. Carmichael. 2004. Mercury in the atmosphere around Japan, Korea, and China as observed during the 2001 ACE-Asia field campaign: measurements, distributions, sources and implications. Journal of Geophysical Research 109(D19S25). Available online at: *http://www.agu.org/journals/jd/jd0416/2003JD004244/2003JD004244.pdf.*

Holloway, T., A. Fiore, and M.G. Hastings. 2003. Intercontinental transport of air pollution: will emerging science lead to a new hemispheric treaty? Environmental Science and Technology 37(20): 4535–4542.

Martin, H. 2003. The Long-Range Transport of Toxics into the Great Lakes Basin: Current Scientific Understanding and Future Priorities. Prepared for the Great Lakes Binational Toxics Strategy Workshop Long-Range Transport Initiative, Ann Arbor, Michigan, September 16–17, 2003.

Pacyna, J.M., E.G. Pacyna, F. Steenhuisen, and S. Wilson. 2002. Mapping 1995 global anthropogenic emissions of mercury. Atmospheric Environment 37(Suppl. 1): S109–SA117.

Sampling and Analysis

Monitoring and Assessing Particulate Matter

JUDITH C. CHOW and JOHN G. WATSON
Desert Research Institute
University of Nevada

Suspended particulate matter (PM) is a pollutant that adversely affects health, regional haze, global climate, ecosystems, and property. Mass measurements of PM were first made in the late nineteenth century by drawing ambient air through filter paper that was weighed before and after sampling. Measurements of total suspended particulates (TSP, particles with an aerodynamic diameter of less than 40 μm) for radioactivity analyses were first done in the United States during the 1950s to evaluate the effects of above-ground nuclear weapons tests. The technology was soon adopted for measuring mass concentrations in major urban areas. The availability of these data were the basis for designating TSP as the first indicator of PM for the 1970 U.S. national ambient air quality standard (NAAQS).

Different particle size fractions, such as PM_{10} and $PM_{2.5}$ (particles with aerodynamic diameters of less than 10 μm and 2.5 μm, respectively) have different properties with respect to penetration into the human respiratory system, effects on visibility, and deposition on surfaces. The composition of PM, mostly carbon, sulfate, nitrate, ammonium, soil-related elements with associated oxides, industrial metals, and liquid water, also influences PM effects (Chow, 1995; Watson et al., 1995).

PM is both emitted directly and formed from gases, such as sulfur dioxide (SO_2), oxides of nitrogen (NO_x), ammonia (NH_3), and volatile organic compounds (VOCs), that are emitted directly into the atmosphere. PM that is emitted directly is called primary aerosol; PM formed from gases is called secondary aerosol. Anthropogenic sources of PM and its precursor gases include industrial emissions, transportation, cooking, heating, the burning of vegetation, and dust from disturbed land. Natural sources include windblown dust from undisturbed deserts, natural vegetation, wildfires, sea salt, and volcanoes. Human activities and industrial development influence both anthropogenic and natural emissions.

Secondary nitrate is related to ozone that derives from NO_x and VOC precursor gases. A large quantity of precursor gases and primary particles comes from combustion sources that also emit carbon dioxide, a greenhouse gas. The problem of high PM concentrations must be addressed in the context of other air pollution problems.

Measuring PM concentrations, identifying their sources, and evaluating the effects of emission-reduction measures are difficult under any circumstances, but they present major challenges in China, where many small industries are not equipped with pollution-control devices and may not be documented in an emissions inventory. In addition, many urban areas use an unknown combination of domestic coal (honeycomb), bottled gas, and natural gas combustion for cooking and heating. Agricultural burning and refuse incineration are also common, even in areas where they have been outlawed.

Environmental planners and policy makers in China must first define the issues related to PM and precursor emissions. They must then adopt valid technical methods to make cost-effective decisions about where to locate roads and facilities, how to implement regulations and pollution-control measures, and where and when to monitor pollution levels to determine hazards, quantify source contributions, and track improvements.

This paper summarizes several approaches to common problems and provides references to more detailed information. Several key topics are addressed as they relate to rapid industrial development in China:

- approaches to designing a PM monitoring network
- zones of influence of different source emissions and zones of representation for ambient monitors
- methods of obtaining and validating PM measurements
- methods of quantifying organic and elemental carbon

DESIGNING A MONITORING NETWORK

The U.S. Environmental Protection Agency provides guidance on locating stations to monitor compliance with NAAQS (Koch and Rector, 1987; Watson et al., 1998). These guidance documents define the concepts and terms of network design and describe methods of defining planning areas and evaluating and selecting monitoring sites for an ambient air-quality network. In addition, the documents suggest how existing resources can be used for network design and provide examples of practical applications of various methodologies.

The first step in designing a network is to define monitoring objectives. Objectives may include: determining compliance with standards; forecasting pollution episodes; and quantifying adverse health effects, source attribution, visibility intensity and causes, materials soiling, and climate change. The next step is site selection. Sites can be selected to represent source influence, community exposure, boundary/transport, and background conditions. Next, the monitoring strategy must be specified. There are three levels of complexity for monitoring networks (EPA, 2002):

- Level III. A network of portable, inexpensive filters for continuous sampling at a large number of locations requires only a small investment in site infrastructure and maintenance. The hardware used for indoor and exposure studies can be used for these sites. The trade-off for greater spatial coverage is reduced accuracy and precision. Temporary, dense networks of this type surrounding Level I and Level II sites could establish zones of representation for the permanent monitors.
- Level II. The network would have fixed sites with proven technology, similar to compliance sites, but with locations and observables that would serve multiple purposes. Sites would have real-time access for forecasting and providing pollution alerts. Resources no longer needed for compliance at urban sites could be used to establish background, boundary, and transport sites. Discontinued Level II compliance sites could be replaced with Level III monitors to address community concerns and issues of environmental justice.
- Level I. The fixed sites in this network would have both proven and novel technologies, similar to U.S. supersites (EPA, 1998). A few of these sites would be located in contrasting environments—with different sources, meteorological conditions, and PM composition—to test new technologies, study atmospheric processes, and support health studies. These sites would have instrumentation similar to that of Level III and Level II sites to determine comparability, as well as detailed particle size ranges, PM chemistry, and precursor gases. They would provide an infrastructure for testing and evaluating new measurement concepts and developing procedures to implement them at Level II and III sites.

ZONES OF INFLUENCE OF SOURCE EMISSIONS AND ZONES OF REPRESENTATION FOR AMBIENT MONITORS

Multiple proximate and distant sources contribute to PM concentrations at any given location. Each monitoring site has a specific zone of representation, depending on the relative amounts contributed by emission sources of different spatial scales. In the United States, these scales are defined in regulation 40 C.F.R., Part 58. Sites outside of urban areas and away from sources provide background monitoring at regional scales (100 km to 1,000 km); sites next to roadways or downwind of factories monitor PM at middle scales (100 m to 500 m) superimposed on contributions from neighborhood (500 m to 4 km), urban (4 km to 100 km), and regional scales.

To estimate the spatial and temporal distribution of PM_{10} concentrations, especially near fugitive dust sources, Chow et al. (1999) set up a closely spaced monitoring network with 29 Level III neighborhood-scale satellite sites surrounding two Level II urban sites with a study domain of 12 km × 13 km. The resultant data showed that the zone of influence around an individual emitter (e.g., a construction site) was less than 1 km. Measurements taken in residential and commercial areas showed that PM_{10} concentrations were similar in neighborhoods.

Another method of determining zones of representation is to examine samples from high time-resolution (e.g., 1 to 5 minute average) measurements. These are available using inexpensive nephelometers, which measure particle light scattering, and aethalometers, which measure particle light absorption. Watson and Chow (2002a) showed how data from neighborhood-scale and urban-scale sites can be compared by subtracting the short-duration spikes that represent middle-scale sources. High time-resolution measurements can also be used to create pollution roses, which can be used to determine the direction from which most of the material originates.

Figure 1 shows how light-scattering measurements taken at a Level I

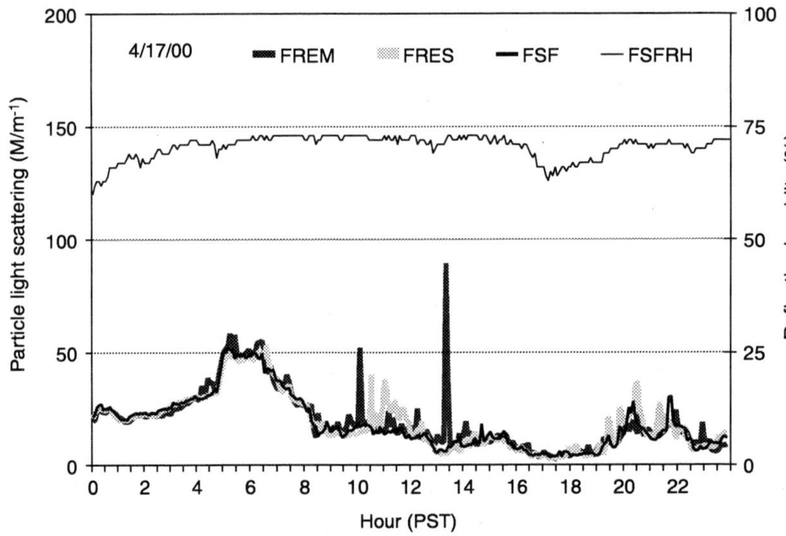

FIGURE 1 Light-scattering measurement taken at the Fresno, California, supersite (FSF), at a nearby residential site (FRES, 0.5 km E of FSF), and at a roadside site (FREM, 1 km WSW of FSF). The relative humidity at FSF (FSFRH) is kept at less than 74 percent by heating the sample inlet when ambient relative humidity exceeds 74 percent. This removes liquid water that would interfere with the relationship between $PM_{2.5}$ and light scattering.

monitoring site (FSF, Fresno, California, supersite) represents the same PM concentrations as were measured in the same neighborhood at a Level III site near a highway (FREM, ~1 km WSW of site FSF) and at a Level III urban residential site (FRES, ~0.5 km east of FSF) (Watson et al., 2000). Neighborhood measurements at FSF show elevated concentrations in the early morning (~0600 hours) and late evening (~2000 hours). However, Figure 1 also shows multiple high-concentration spikes (mostly from the roadside site and typically during periods of heavy traffic) that were not observed at the Fresno supersite (FSF).

OBTAINING AND VALIDATING MEASUREMENTS

Several reviews identify the challenges for PM sampling and analysis (e.g., Chow, 1995; Chow et al., 2002a; EPA, 1994; Watson and Chow 2001; Wilson et al., 2002). The 24-hour, integrated, filter-based technology is gradually evolving toward in-situ, continuous, hourly measurements. In the interim, precision, equivalence, comparability, and predictability must be established among different measurement principles and methods. Short-duration (3 to 6 hours), rather than 24-hour, integrated measurements, are often used to monitor diurnal variations of PM concentrations and composition. Flow rates for short-duration samples must be increased to ensure adequate loadings for chemical speciation.

No single filter medium is appropriate for all chemical analyses; chemical characterization often requires sampling of multiple substrates (Chow, 1995; Fehsenfeld et al., 2003). Selection of the appropriate sampling substrates depends on particulate sampling efficiency, mechanical stability, chemical stability, temperature stability, and cost/availability. Because Teflon-membrane filters are inert to the absorption of gases, do not absorb water, and have low blank weights and low blank levels, they have been commonly used for mass measurement by gravimetry and elemental measurement by x-ray fluorescence (XRF), photo-induced x-ray emission (PIXE), and instrumental neutron-activation analysis (INAA) (Watson et al., 1998). Prewashed and prebaked quartz-fiber filters, which have low blank levels for ions and carbon and do not absorb water, are used for anion and cation analysis by ion chromatography (IC) (Chow and Watson, 1998) and carbon analysis (Chow et al., 1993, 2001). Several sampling substrates are subject to positive and negative interferences, such as artifact formation, loss of volatile species, filter fragility, particle loss in transport, and water absorption when the relative humidity is high.

Chow et al. (2002b) showed that 20 to 30 percent of sampled ammonium nitrate may evaporate if left in the field for 24 hours; 40 to 50 percent is volatilized if the sample is left in the field for 72 hours. Volatilized nitrate is not significant (within ± 5 percent) with diurnal monitoring in the early morning (0000 to 0600 hours) and late evening (1800 to 2400 hours). However, $PM_{2.5}$ volatilized nitrate may be significant (~20 to 30 percent during the morning hours and up to ~ 40 percent during the afternoon hours when temperatures exceed

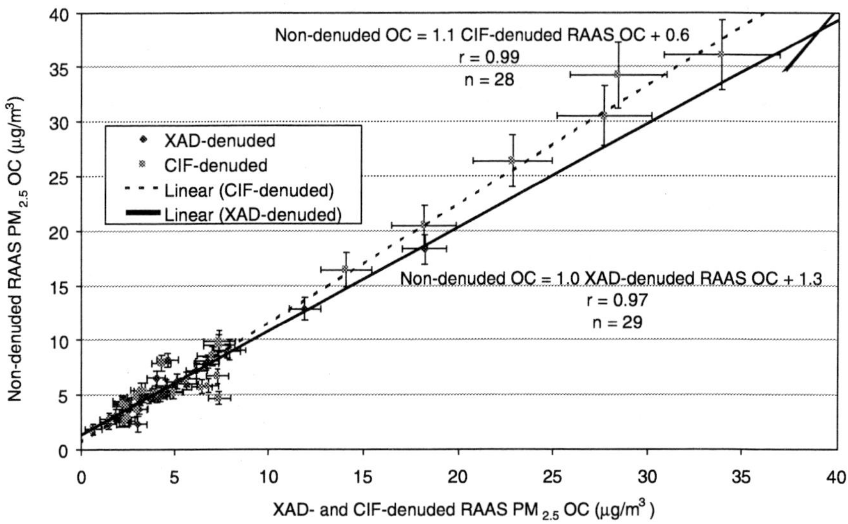

FIGURE 2 Measurements of denuded and nondenuded organic carbon (OC) are similar when field blanks are subtracted (CY2000). Denuders consisting of XAD resin or charcoal-impregnated cellulose-fiber filter (CIF) remove organic vapors from the incoming airstream.

20°C). Because volatilized ammonium nitrate is not measured as part of the gravimetric mass, $PM_{2.5}$ mass concentrations may be underestimated.

Denuders and backup filters are often used to evaluate and quantify positive and negative artifacts. Denuders remove gases that might absorb onto a filter; backup filters capture gases that have evaporated from sampled particles. Positive and negative artifacts may partially cancel each other out in measurements of organic carbon. At the FSF site, for example, Watson and Chow (2002b) found that organic carbon samples were similar with and without preceding organic-gas denuders when a field blank was subtracted (Figure 2). Denuded and nondenuded samples of organic carbon were highly correlated ($0.97 < r < 0.99$) with a slope close to unity (within ± 10 percent). Detailed measurements of organic compounds coupled with accurate thermodynamic models that explain gas particle-phase relationships are necessary for measuring different aerosol compositions in various environments.

QUANTIFYING ORGANIC AND ELEMENTAL CARBON

Carbon, a large component of PM (30 to 40 percent), is often divided into several fractions, most commonly organic carbon and elemental carbon. Elemental carbon, the black, light-absorbing fraction, is measured by many different methods that do not always have the same results (Chow et al., 1993, 2001; Currie at al., 2002; Fung et al., 2002; Schmid et al., 2001; Sharma et al., 2002; Watson and Chow, 2002b). Some of the more detailed carbon fractions have been found to represent different source contributions (Chow et al., 2003, 2004a; Chow and Watson, 2002; Watson et al., 1994).

Fung et al. (2002) compared the interagency monitoring of protected visual environments (IMPROVE) thermal optical reflectance (TOR) method (Chow et al., 1993) with the thermal manganese dioxide oxidation (TMO) method (Fung, 1990) for simulated black carbon and ambient samples from Hong Kong, Korea, and the United States. The results for elemental carbon compared well with the simulated sample, but the TMO measurement was higher than the IMPROVE TOR measurement for the samples in urban, vehicle-dominated Hong Kong and lower than the IMPROVE TOR measurement in the remote background Korean and U.S. samples. This comparison showed that different carbon analysis methods can vary markedly as a function of the composition and concentration of the sample.

Chow et al. (2004b) showed that elemental carbon differs within a temperature protocol for reflectance (TOR) and transmittance (TOT) pyrolysis corrections, as shown in Figure 3. The Environmental Protection Agency (EPA) uses two different protocols for thermal/carbon analysis: (1) the IMPROVE network protocol and (2) the speciation trends network (STN) protocol. Both protocols sample air through quartz-fiber filters. Punches from these filters are then submitted for analysis (Chow et al., 1993; 2001). The two methods differ in terms of combustion temperatures, analysis durations, and optical monitoring. As shown in Figure 3, the underestimation of elemental carbon is ~30 percent with the IMPROVE monitoring protocol; this increases to ~70 percent with the STN protocol. When the cross-sections of the filter punches were examined: (1) charring was found throughout the filter, either from adsorbed organic vapors during sampling or adsorbed vaporized particles during analysis or liquid organic particles; (2) elemental carbon near the filter surface evolves earlier than some of the charred organic carbon embedded in the filters during thermal analysis; and (3) charred organic carbon attenuates red laser light more efficiently than elemental carbon (Chow et al., 2004b).

Many operating parameters can individually or collectively cause variations of up to 30 percent in test results. These include differences in combustion atmospheres, temperature ramping rates, and temperature plateaus. Unfortunately, no systematic research is being done to resolve these discrepancies. Such research should include: (1) a review and evaluation of the published literature, much of which is not in the air-quality journals; (2) documentation of the method used, especially for thermal evolution carbon comparisons with different combustion temperatures, residence times, ramp rates, and optical-pyrolysis corrections; (3) establishment of standards for different sources of black carbon (e.g., diesel fuel, wood burning, tar, gasoline engines) on different filter media (e.g., Teflon

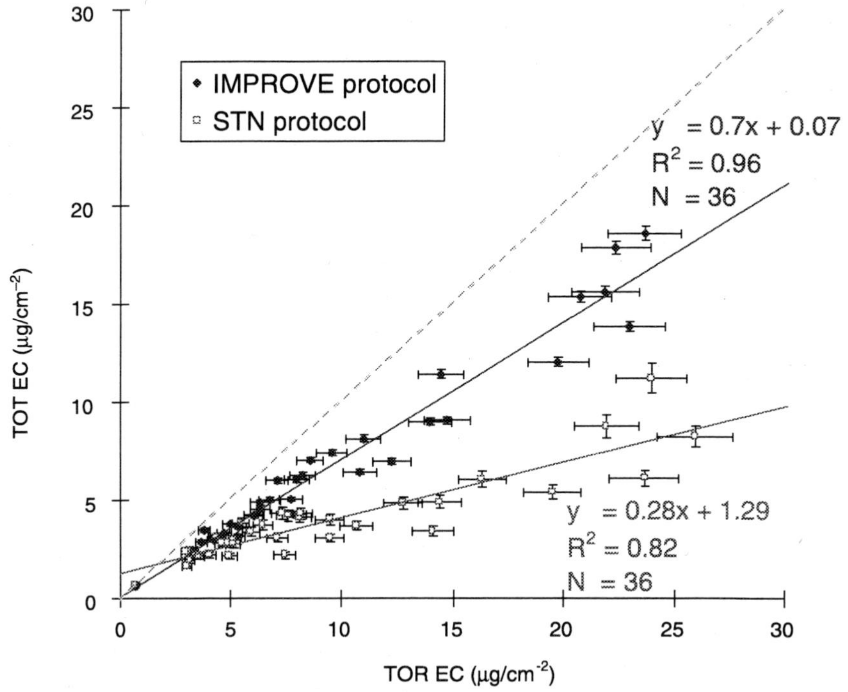

FIGURE 3 Comparison of transmission (TOT) and reflectance (TOR) corrected measurements of elemental carbon (EC) in samples from Fresno, California, using the IMPROVE and STN protocols. (The dashed line indicates the 1:1 correspondence.) Source: Chow et al., 2004b.

membranes, quartz fibers) at different deposit levels (light, medium, and dark appearances) for measurements of absolute absorption with a photoacoustic monitor (Moosmüller et al., 1998); (4) optical modeling of changes in absorption properties of particles on and within a filter compared to particles suspended in air (e.g., Chen et al., 2004); (5) experiments to examine the distribution of particles and pyrolysis artifact in a filter; (6) an evaluation of the effects of nonabsorbing particles, transmittance wavelengths, initial darkness, carbonate deposits, and oxygen-supplying minerals; and (7) optimizing fractions to identify source contributions.

RELEVANCE TO CHINA

China is a large, diverse country with many local, urban, regional, and international pollution problems for which there are no simple solutions. In this paper we have presented some possible starting points for PM monitoring and assessments.

Because of China's rapid urbanization, pollution control will require collaboration among academic, regulatory, and commercial entities in various cities and provinces, and across international borders.

An integrated assessment, similar to the North American Research Strategy for Tropospheric Ozone (NARSTO) public/private partnership (Fehsenfeld et al., 2003), would be a good starting point for determining what is known and unknown about PM in China and what steps should be taken to minimize its adverse effects on health, visibility, and the environment.

REFERENCES

Chen, L.-W.A., J.C. Chow, W.P. Arnott, H. Moosmüller, and J.G. Watson. 2004. Modeling reflectance and transmittance of quartz-fiber filter samples containing elemental carbon particles: implications for thermal/optical analysis. Journal of Aerosol Science. In press.

Chow, J.C. 1995. Measurement methods to determine compliance with ambient air quality standards for suspended particles. Journal of the Air and Waste Management Association 45(5): 320–382.

Chow, J.C., and J.G. Watson. 1998. Ion Chromatography in Elemental Analysis of Airborne Particles. Pp. 97–137 in Elemental Analysis of Airborne Particles, Vol. 1, edited by S. Landsberger and M. Creatchman. Amsterdam: Gordon and Breach Science.

Chow, J.C., and J.C. Watson. 2002. Review of $PM_{2.5}$ and PM_{10} apportionment for fossil fuel combustion and other sources by the chemical mass balance receptor model. Energy and Fuels 16(2): 222–260. Available online at: *http://pubs3.acs.org/acs/journals/doilookup?in_doi=10.1021/ef0101715.*

Chow, J.C., J.G. Watson, L.C. Pritchett, W.R. Pierson, C.A. Frazier, and R.G. Purcell. 1993. The DRI thermal/optical reflectance carbon analysis system: description, evaluation and applications in U.S. air quality studies. Atmospheric Environment 27A(8): 1185–1201.

Chow, J.C., J.G. Watson, M.C. Green, D.H. Lowenthal, D.W. DuBois, S.D. Kohl, R.T. Egami, J.A. Gillies, C.F. Rogers, C.A. Frazier, and W. Cates. 1999. Middle- and neighborhood-scale variations of PM_{10} source contributions in Las Vegas, Nevada. Journal of the Air and Waste Management Association 49(6): 641–654.

Chow, J.C., J.G. Watson, D. Crow, D.H. Lowenthal, and T. Merrifield, T. 2001. Comparison of IMPROVE and NIOSH carbon measurements. Aerosol Science and Technology 34(1): 23–34.

Chow, J.C., J.P. Engelbrecht, J.G. Watson, W.E. Wilson, N.H. Frank, and T. Zhu. 2002a. Designing monitoring networks to represent outdoor human exposure. Chemosphere 49(9): 961–978.

Chow, J.C., J.G. Watson, S.A. Edgerton, and E. Vega. 2002b. Chemical composition of $PM_{2.5}$ and PM_{10} in Mexico City during winter 1997. Science and the Total Environment 287(3): 177–201.

Chow, J.C., J.G. Watson, L.L. Ashbaugh, and K.L. Magliano. 2003. Similarities and differences in PM_{10} chemical source profiles for geological dust from the San Joaquin Valley, California. Atmosphere and Environment 37(9-10): 1317–1340.

Chow, J.C., J.G. Watson, H.D. Kuhns, V. Etyemezian, D.H. Lowenthal, D.J. Crow, S.D. Kohl, J.P. Engelbrecht, and M.C. Green. 2004a. Source profiles for industrial, mobile, and area sources in the Big Bend Regional Aerosol Visibility and Observational Study. Chemosphere 54: 185–208.

Chow, J.C., J.G. Watson, L.-W.A. Chen, W.P. Arnott, H. Moosmüller, and K. Fung. 2004b. Equivalence of elemental carbon by thermal/optical reflectance and transmittance with different temperature protocols. Environmental Science and Technology. Accepted.

Currie, L.A., B.A. Benner, Jr., H. Cachier, R. Cary, J.C. Chow, E.R.M. Druffel, T.I. Eglinton, O. Gustafsson, P.C Hartmann, J.I. Hedges, J.D. Kessler, T.W. Kirchstetter, D.B. Klinedinst, G.A. Klouda, J.V. Marolf, C.A. Masiello, T. Novakov, A. Pearson, K.M. Prentice, H. Puxbaum, J.G. Quinn, C.M. Reddy, H. Schmid, J.F. Slater, J.G. Watson, and S.A. Wise. 2002. A critical evaluation of interlaboratory data on total, elemental, and isotopic carbon in the carbonaceous particle reference material, NIST SRM 1649a. Journal of Research of the National Institute of Standards and Technology 107(3): 279–298.

EPA (Environmental Protection Agency). 1994. Guidelines for PM_{10} Sampling and Analysis Applicable to Receptor Modeling, Final Report. EPA-452/R-94-009. Research Triangle Park, N.C.: Environmental Protection Agency, Office of Air Quality Planning and Standards.

EPA. 1998. Draft Supersites Conceptual Plan. Research Triangle Park, N.C.: Environmental Protection Agency, Office of Air Quality Planning and Standards.

EPA. 2002. National Air Quality Monitoring Strategy. Available online at: *http:// www.epa.gov/ttn/amtic/stratdoc.html.*

Fehsenfeld, F.C., D.R. Hastie, P.A. Solomon, and J.C. Chow. 2003. Gas and Particle Measurments. Pp. 5-1–5-37 in Particulate Matter Science for Policy Makers: A NARSTO Assessment, Part 2. Pasco, Wash.: NARSTO. Also available online at: *http://www.cgenv.com/Narsto.*

Fung, K.K. 1990. Particulate carbon speciation by MnO_2 oxidation. Aerosol Science and Technology 12(1): 122–127.

Fung, K.K., J.C. Chow, and J.G. Watson. 2002. Evaluation of OC/EC speciation by thermal manganese dioxide oxidation and the IMPROVE method. Journal of the Air and Waste Management Association 52(11): 1333–1341.

Koch, R.C., and H.E. Rector. 1987. Optimum network design and site exposure criteria for particulate matter. EPA 450/4-87-009. Research Triangle Park, N.C.: Environmental Protection Agency, Office of Air Quality Planning and Standards.

Moosmüller, H., W.P. Arnott, C.F. Rogers, J.C. Chow, C.A. Frazier, L.E. Sherman, and D.L. Dietrich. 1998. Photoacoustic and filter measurements related to aerosol light absorption during the Northern Front Range Air Quality Study (Colorado 1996/1997). Journal of Geophysical Research 103(D21): 28149–28157.

Schmid, H.P., L. Laskus, H.J. Abraham, U. Baltensperger, V.M.H. Lavanchy, M. Bizjak, P. Burba, H. Cachier, D.J. Crow, J.C. Chow, T. Gnauk, A. Even, H.M. ten Brink, K.P. Giesen, R. Hitzenberger, C. Hueglin, W. Maenhaut, C.A. Pio, and J. Puttock. 2001. Results of the "Carbon Conference" international aerosol carbon round robin test: Stage 1. Atmospheric Environment 35(12): 2111–2121.

Sharma, S., J.R. Brook, H. Cachier, J.C. Chow, A. Gaudenzi, and G. Lu. 2002. Light absorption and thermal measurements of black carbon in different regions of Canada. Journal of Geophysical Research 109(D24): AAC 11-1–AAC 11-11.

Watson, J.G., and J.C. Chow. 2001. Ambient Air Sampling. Pp. 821–844 in Aerosol Measurement: Principles, Techniques, and Applications, 2nd ed, edited by P. Baron and K. Willeke. New York: John Wiley & Sons.

Watson, J.G., and J.C. Chow. 2002a. A wintertime $PM_{2.5}$ episode at the Fresno, CA supersite. Atmospheric Environment 36(3): 465–475.

Watson, J.G., and J.C. Chow. 2002b. Comparison and evaluation of in situ and filter carbon measurements at the Fresno supersite. Journal of Geophysical Research 107(D21): ICC 3-1–ICC 3-15.

Watson, J.G., J.C. Chow, D.H. Lowenthal, L.C. Pritchett, C.A. Frazier, G.R. Neuroth, and R. Robbins. 1994. Differences in the carbon composition of source profiles for diesel- and gasoline-powered vehicles. Atmosphere and Environment 28(15): 2493–2505.

Watson, J.G., G.D. Thurston, N.H. Frank, J.P. Lodge, R.W. Wiener, F.F. McElroy, M.T. Kleinman, P.K. Mueller, and J.C. Chow. 1995. Measurement methods to determine compliance with ambient air quality standards for suspended particles: 1995 critical review discussion. Journal of the Air and Waste Management Association 45(9): 666–684.

Watson, J.G., J.C. Chow, H. Moosmuller, M.C. Green, N.H. Frank, and M.L. Pitchford. 1998. Guidance for using continuous monitors in $PM_{2.5}$ monitoring networks. EPA-454/R-98-012. Research Triangle Park, N.C.: Environmental Protection Agency, Office of Air Quality Planning and Standards.

Watson, J.G., J. C. Chow, and C.A. Frazier. 1999. X-ray Fluorescence Analysis of Ambient Air Samples. Pp. 67–96 in Elemental Analysis of Airborne Particles, Vol. 1, edited by S. Landsberger and M. Creatchman. Amsterdam: Gordon and Breach Science.

Watson, J.G.,s J.C. Chow, J.L. Bowen, D.H. Lowenthal, S.V. Hering, P. Ouchida, and W. Oslund. 2000. Air quality measurements from the Fresno supersite. Journal of the Air and Waste Management Association 50(8): 1321–1334.

Wilson, W.E., J.C. Chow, C.S. Claiborn, W. Fusheng, J.P. Engelbrecht, and J.G. Watson. 2002. Monitoring of particulate matter outdoors. Chemosphere 49(9): 1009–1043.

Source Apportionment of Fine-Particle Pollution in Beijing

YUANHANG ZHANG, XIANLEI ZHU, and LIMIN ZENG
College of Environmental Sciences
Peking University

WEI WANG
China Research Academy of Environmental Sciences

Airborne particulates, especially fine particulates, have serious effects on visibility, climate, and human health. Fine particles can penetrate the human respiratory tract and lungs, and several epidemiological studies have reported a link between elevated particle concentration and increased mortality and morbidity (e.g., Abbey et al., 1998; Ostro et al., 1999; Wilson and Suh, 1997). In addition, aerosols influence both atmospheric visibility and climate through the scattering and absorption of solar radiation (Schwartz, 1996).

China, which is undergoing rapid economic development and population growth, is facing serious and complicated air pollution problems that have resulted in a unique chemical transformation and transport process. In the past several decades, air pollution in China was caused mainly by the burning of coal as fuel for industrial and domestic purposes; sulfur dioxide (SO_2) and particulate matter were among the major pollutants. Unfortunately, owing to rapid urbanization and industrialization, before the existing problems caused by coal combustion could be resolved, other emission sources have become increasingly important. The rapid increase in the number of vehicles in some Chinese megacities and economically developed regions, such as Beijing, Shanghai, Guangzhou, Pearl River delta, and Yangtze delta, has led to a sharp increase in concentrations of nitrogen oxides (NO_x), volatile organic compounds (VOCs), particulate matter (PM), and ozone (Tang et al., 1995; Zhang et al., 1997, 1998). The combination of coal smog and traffic exhaust results in serious pollution characterized by high levels of photochemical smog, high concentration of fine particles, and poor visibility. Air pollution of this kind causes both local and regional problems.

Beijing, the capital of the People's Republic of China, is located on the North China Plain at an elevation of 44 meters above mean sea level. Beijing is a typical Chinese megacity with a population of more than 11 million. The city, which has undergone rapid development since the 1980s, has high concentrations of particulate matter and poor visibility in spite of the adoption of numerous measures to control particle pollution. Measured annual mean mass concentrations of fine particles (PM_{10}) exceed the Grade II (100 $\mu g/m^3$) National Ambient Air Quality Standard (NAAQS) (Song et al., 2002a,b). Concentrations of $PM_{2.5}$, particles with aerodynamic diameters of less than 2.5 $\mu g/m^3$, are also much higher than the recommended standard for annual average ground-level $PM_{2.5}$ in the United States (15 $\mu g/m^3$). The ratio of $PM_{2.5}$ concentration to PM_{10} concentration ($PM_{2.5}/PM_{10}$ ratio) ranges from 0.5 to 0.7, with an average of 0.6, which is about the same as the ratio observed in Europe and the United States. Measurements of $PM_{2.5}$ performed along wind direction suggest that anthropogenic pollution in urban areas extends to a regional scale under some meteorological conditions and that the whole Beijing-Tianjin area is sometimes covered with a large polluted air mass. Studies in Beijing showed an inverse correlation between elevated concentrations of $PM_{2.5}$ and visibility based on hourly measurements in June 1999 and January 2000 (Bergin et al., 2001).

High levels of particulate matter and adverse effects have inevitably increased concerns about how fine particles can be controlled. As a basis for developing effective control strategies for fine-particle pollution and improving air quality in Beijing, we must first determine the relative importance of the various sources that contribute to $PM_{2.5}$. The goals of this study were: (1) to quantify the source contributions to $PM_{2.5}$ in Beijing by source inventory and chemical mass balance (CMB) model; (2) to compare the results of these two methods; and (3) to investigate the spatial and seasonal variations of source contributions.

EXPERIMENTAL METHODS

Sampling Sites

The samples of airborne $PM_{2.5}$ used in this study were collected at three sites: College of Chemistry of Beijing Union University (BUU), Chinese Academy of Preventive Medicine (CAPM), and Chinese Research Academy of Environmental Sciences (CRAES).

The College of Chemistry of BUU, located at Fatou town in the southeast of Beijing, is 2 kilometers from the Eastern Fourth Ring Road and 500 meters west of a chemical-industry zone. The sampling location was on the roof of the Main Teaching Building; there were no high-rise buildings nearby. The two roads, to the east and north, had little traffic.

CAPM, situated in the south downtown area of Beijing, is west of the Eastern Secondary Ring Road and 300 meters from Guangming Bridge. Sampling instruments were placed on the southern side of the roof of the three-story office building. The sampling site is adjacent to Panjiayuan East Street (to the east), with high-rise buildings to the south and north.

CRAES is located in the northeast suburbs of Beijing, about 100 meters north of the Lishui Bridge. The sampling location was on the northern side of the roof of the four-story Science and Research Building. The building is 100 meters east of Beiyuan Street, a busy main roadway connecting the urban area with Changping and the residential area in the northern part of the city. Trees between the building and the street decrease the effects of vehicle exhaust.

Ambient Sampling and Analysis

Atmospheric samples were taken at the three sites during April 25–30, 2000 (spring); August 18–25, 2000 (summer); October 30–November 4, 2000 (autumn); and January 9–14, 2001 (winter). Because there were not enough identical instruments for sampling in the three sites simultaneously, different instruments were used, including a MOUDI-100 impactor (MSP Company), an A-245 dichotomous sampler (CIREE Company of USA), a $PM_{2.5}$ sampler (Beijing Geological Instrument Factory), and a self-developed sampler. Teflon filters and quartz-fiber filters were used to collect samples to determine chemical species.

The meteorological conditions varied during the four sampling periods. On April 25, 2000, air quality had deteriorated because of a sandstorm, but sunny days followed. In the summer sampling period, the weather was mostly rainy or cloudy, unlike the typical summer weather, which is characterized by high temperatures and strong sunshine. During November 1–4, the weather was mild and winds were light. During the January sampling period, it snowed but then remained sunny or partly sunny.

The chemical analysis for ambient samples has been described extensively elsewhere (Tang, 2001). Briefly, inorganic elements and ions in $PM_{2.5}$ collected by teflon filters were quantified by induced, coupled, plasma-atomic emission spectrometry (ICP-AES) and ion chromatography (IC), respectively. Organic and elemental carbons (OC and EC) in $PM_{2.5}$ collected on quartz-fiber filters were determined by means of optical/thermal techniques. Organic compounds, including polycyclic aromatic hydrocarbons (PAHs), were determined by gas chromatography/mass spectrometry (GC/MS).

Source Tests and Source Profiles

The source emission profiles used in the present study were obtained from source tests in Beijing and literature that provided emission rates of OC and EC, particle-phase PAHs, elements, and ions for the major sources (Table 1) (Hildemann et al., 1991; Watson et al., 2001). Soil dust, road dust, and wind-blown dust, which have similar profiles, were combined as fugitive dust.

TABLE 1 Source Profiles of $PM_{2.5}$ (μg/g)

	Coal Combustion	Vehicle Exhaust	Construction Dust	Fugitive Dust	Biomass Burning	Secondary Sulfates	Secondary Nitrates
Fluoranthene	1,186	2,828.4[a]	157.33	44.66	0.0	0.0	0.0
Pyrene	490	6,512.3[a]	93.20	29.58	0.0	0.0	0.0
Benz[a]anthracene	1,083	645.31	27.02	9.81	0.0	0.0	0.0
Chrysene	1,794	1,053.1	66.15	26.78	0.0	0.0	0.0
Benzo [k+b]fluoranthene	3,755	456.66	93.00	34.22	0.0	0.0	0.0
Benzo[e]pyrene	627	504.17	84.13	30.44	0.0	0.0	0.0
Benzo[a]pyrene	1,048	617.05	31.68	8.34	0.0	0.0	0.0
Perylene	365	85.03	3.38	2.36	0.0	0.0	0.0
Indeno [1,2,3-cd]pyrene	1,061	223.01	28.07	8.84	0.0	0.0	0.0
Benzo [ghi]perylene	836	232.61	30.45	9.40	0.0	0.0	0.0
Dibenzo [a,h]anthracene	240[a]	39.96	1.12	0.27	0.0	0.0	0.0
Organic carbon	24,800	390,000	0.0	186,800	484,000	0.0	0.0
Elemental carbon	10,700	365,000[a]	0.0	15,700	28,600	0.0	0.0
Aluminum	47,800[a]	2,160	42,600	73,100[a]	110	0.0	0.0
Calcium	4,400	3,373	300,000[a]	46,500	100	0.0	0.0
Potassium	11,000	2,111	16,100	12,100	16,700[a]	0.0	0.0
Nitrate	3,147	64,400	0.0	1,100	2,500	0.0	775,000[a]
Sulfate	263,600	28,000	0.0	11,000	2,500	727,000[a]	0.0
Ammonium	200	15,500	0.0	100	1,500	273,000	226,000

[a]Tracer.

Sources: Bergin et al., 2001; Hildemann et al., 1991; Tang, 2001.

Documented emission rates of potassium (K), OC and EC, aluminum (Al), calcium (Ca), nitrate (NO_3^-), sulfate (SO_4^{2-}), and ammonium (NH_4^+) for wood combustion (Hildemann et al., 1991) were used as the source profile for biomass burning because no study of fine-particle emissions from crop straw burning in Beijing was available. The source profiles of secondary sulfate and nitrate, which are formed in the atmosphere, were assumed to be ammonium sulfate and ammonium nitrate, respectively.

METHODOLOGY

Three approaches can be used to evaluate source contributions to airborne particles: source inventory; dispersion models; and receptor models.

The source inventory method is based on a compilation of measured emissions from given source types and an integration of all activities by those sources in the studied region. The method of constructing a source inventory is well defined, and source inventories are widely available in many countries. However, some components of particulate matter, such as re-suspended soil and dust, are extremely difficult to quantify accurately and are subject to seasonal variations that are not well understood. Thus, although source inventories can be used confidently for the major primary sources, such as traffic and power plant emissions, the quantification of fugitive emissions and estimates of secondary sources of sulfate and nitrate formed in the atmosphere are highly problematic. In addition, because emissions from different categories of sources often occur at different heights, their impacts on airborne particle concentrations are not directly proportional to their relative source strengths; this circumstance is not given full consideration in source inventories (Harrison et al., 1997).

Dispersion models use emissions data and transport calculations to predict pollutant concentrations at specific air-monitoring locations (Bencala and Seinfeld, 1979; Liu and Seinfeld, 1975). Unlike receptor models, dispersion models are definitive enough to identify contributions of individual sources within a class. Dispersion models have been used for a long time to develop control strategies for air pollutants. However, as mentioned above, estimations based on dispersion models sometimes contain considerable uncertainty because the inventories they rely upon are often inaccurate for some sources (Gordon, 1980).

Receptor models assess contributions from various sources based on observations at sampling sites, the “receptors” (Gordon, 1980; Watson, 1984). Techniques include CMB and some purely statistical approaches, such as principle component analysis (PCA). CMB requires a detailed knowledge of the composition of emissions from individual sources of all types, which is rarely available in an urban area with many potential and diverse sources. One of the limitations of multivariate-factor analyses is that they require the collection and analysis of large numbers of ambient air samples and statistically independent source tracers for each major source type.

The CMB approach used in this study infers source contributions by determining the best-fit linear combination of emission source profiles to reconstruct the measured chemical composition (elements, ions, OC/EC, PAHs) of ambient $PM_{2.5}$ samples. This can be expressed by the following equation:

$$C_{ik} = \sum_{j=1}^{m} \alpha_{ij} S_{ij}$$

where C_{ik} is the concentration of chemical species i in fine particles at receptor site k; α_{ij} is the relative concentration of chemical constituent i in the fine-particle emissions from source j; and S_{jk} is the mass contributed to total fine particulate at receptor site k originating from source j.

RESULTS AND DISCUSSION

Chemical Composition of $PM_{2.5}$

The complicated chemical composition of fine particles gives some clues about their sources. Elements such as lead, arsenic, vanadium, nickel, and cadmium originate from anthropogenic sources. Crustal elements, such as aluminum, silicon, and iron, often originate in natural sources (e.g., soil dust). Some inorganic ions, such as sulfate and nitrate, are the results of atmospheric transformations of SO_2 and NO_x. EC derives from some primary sources (e.g., combustion). OC derives from both primary sources (e.g., cooking and biomass burning) and the formation of gas-phase organic compounds in the atmosphere.

The typical chemical composition of $PM_{2.5}$ in Beijing is shown in Figure 1. OC is the largest component, accounting for 31.8 percent. Crustal components, sulfate, and nitrate constitute 9.6, 7.2, and 4.7 percent, respectively. Ammonium, EC, and other elements account for relatively small proportions. Compared with some cities in other countries, Beijing has much higher concentrations of OC, sulfate, and nitrate (Castanho and Artaxo, 2001; Harrison et al., 1997; Lin and Tai, 2001; Yantin et al., 2000). This may be because secondary aerosols formed as a result of strong local emissions of VOCs, SO_2, and NO_x and the high oxidation capacity of the atmosphere account for a large fraction of the $PM_{2.5}$ concentrations in Beijing. The high concentration of mineral elements indicates another important source of $PM_{2.5}$—soil dust.

Fine-Particle Sources

Source Contributions from Source Inventories

The pollution sources of $PM_{2.5}$ in Beijing were classified into three categories: stationary sources (e.g., fuel combustion for various purposes); mobile sources (e.g., vehicle exhaust); and fugitive sources (e.g., construction dust, paved

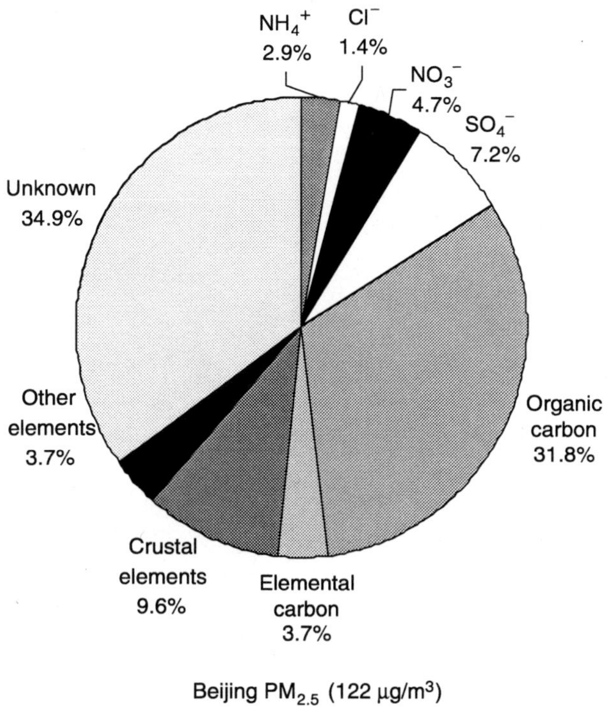

FIGURE 1 Chemical composition of $PM_{2.5}$ in Beijing.

and unpaved road dust, industrial dust, etc.). After sources had been characterized and emissions measured, and using 1999 as a reference year, a source inventory was developed for the eight districts of Beijing (1,040 km^2).

All pollution sources discharge a combined 53,266 tons of $PM_{2.5}$ per year. As shown in Figure 2, stationary sources account for 46.1 percent, mobile sources 15.4 percent, and combined fugitive sources 38.5 percent. Thus, stationary and fugitive sources are the major sources of $PM_{2.5}$ in Beijing. In the stationary source category, point sources and area sources contribute 18.0 percent and 28.2 percent, respectively, to total emissions. In the fugitive source category, road dust and industrial dust contribute 18.7 percent and 12.4 percent, respectively; construction dust and dust from unpaved roads make only minor contributions.

Source Contributions from Chemical Mass Balance Models

Seven major sources of $PM_{2.5}$ are identified by the CMB model (Figure 3): coal combustion; vehicle exhaust; construction dust; fugitive dust; biomass

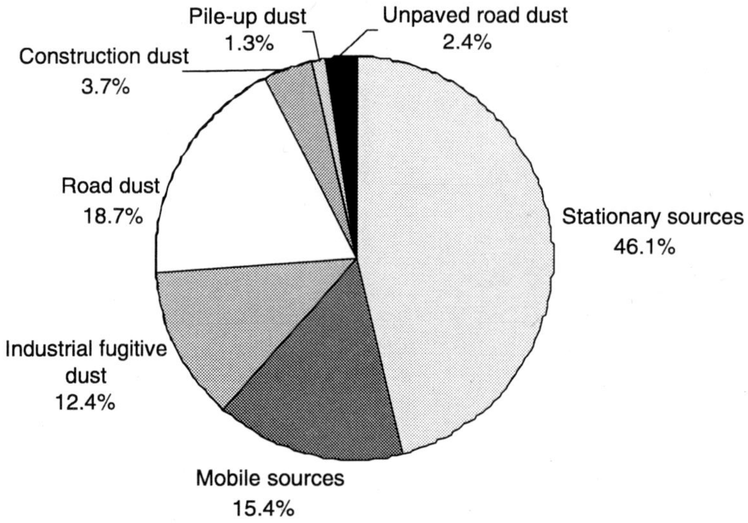

FIGURE 2 Source inventory for $PM_{2.5}$ in Beijing.

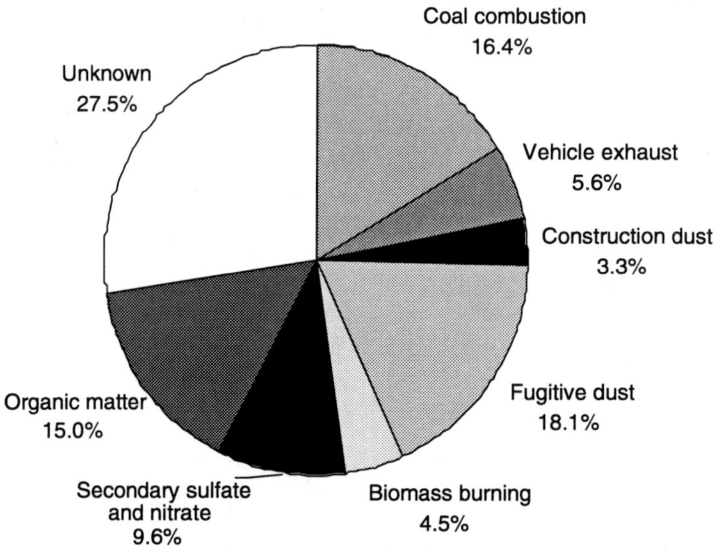

FIGURE 3 Sources of $PM_{2.5}$ in Beijing identified by CMB model.

burning; secondary sulfate and nitrate; and organic matter. These are consistent with the results computed by positive matrix factorization (PMF) (Song et al., 2002a). Together these seven sources contributed 72.5 percent of $PM_{2.5}$ mass concentration.

Fugitive dust and coal combustion, with contributions of 18.1 and 16.4 percent, respectively, are largely responsible for $PM_{2.5}$ in Beijing. The higher contribution of fugitive dust in Beijing than in some other cities overseas is attributable to the dry climate and lack of vegetative coverage (Schauer et al., 1996; Zheng et al., 2002). Because coal is by far the most important fuel in China, the substantial effect of coal combustion on urban air pollution is not surprising. The large contribution of coal combustion in Beijing suggests there is a long way to go to control this source effectively.

To estimate the contribution of vehicle exhaust, PAHs are used as tracers; the PAH source profiles for vehicle exhaust and road dust have less linearity than profiles of inorganic elements. The CMB model shows that the contribution from vehicle emissions is 5.6 percent, much lower than the contribution of 15.5 percent obtained by using the vehicle/road dust source profile of inorganic elements. If the difference between 15.5 and 5.6 percent is a rough estimate of the contribution of road dust, it is apparent that road dust has a significant effect on $PM_{2.5}$. Construction dust and biomass burning contribute a small amount to $PM_{2.5}$, with contributions of 3.3 and 4.5 percent, respectively.

Besides primary sources, about 9.6 percent of $PM_{2.5}$ mass concentration is attributable to secondary formation of sulfate and nitrate in the atmosphere. Organic matter in fine particles derives from both primary sources and organic compounds formed in the atmosphere (i.e., secondary organic aerosol [SOA]). The contribution of 15 percent from organic matter computed by the CMB model from both primary sources and SOA. According to studies, the difference between total fine-particle OC and OC from primary sources, which accounts for 19.2 to 34.1 percent of OC in $PM_{2.5}$, is related to SOA (Schauer and Cass, 2000; Schauer et al., 1996; Zheng et al., 2002). Based on this percentage and measured OC concentrations in Beijing (22.4 percent annual average), the contribution of SOA to $PM_{2.5}$ is about 6.0 to 10.6 percent. The combined contributions of secondary sulfate, secondary nitrate, and SOA is a remarkable 15.6 to 20.2 percent, suggesting that limiting secondary sources might be decisive in improving air quality in Beijing.

Comparison of Results

Table 2 compares the average source contributions calculated by the source inventory and CMB model. Because the source inventory only provides information on primary sources, the contributions of primary sources computed by CMB have been normalized in the last row of the table for easy comparison.

Both the source inventory and the CMB model suggest that the major

TABLE 2 Comparison of Results from the Source Inventory and Chemical Mass Balance Model

	Source Contribution (percentage)		
Source	Source Inventory	CMB	CMB (normalized)[a]
Coal combustion	46.1	16.4	37.8
Mobile sources	15.4	5.6	12.9
Fugitive sources	38.5	21.4	49.3
Industrial dust	12.4	NA	
Road dust	18.7	NA	
Construction dust	3.7	NA	
Pile-up dust	1.3	NA	
Unpaved road dust	2.4	NA	
Biomass burning	NA	4.5	
Secondary sulfate and nitrate	NA	9.6	
Organic matter	NA	15.0	
Unknown	NA	27.5	
Total	100	100	100

[a]Normalization of the contributions from coal combustion, mobile sources, and fugitive sources.

primary sources of $PM_{2.5}$ in Beijing are coal combustion and fugitive dust; mobile sources contribute a minor fraction of $PM_{2.5}$. The two methods also provide comparable contributions of about 13 to 15 percent for vehicle exhaust. However, for coal combustion, the source inventory estimates a higher contribution than the CMB model. This is probably because the source inventory does not take into account the reduced pollution effects on ambient $PM_{2.5}$ from high chimney emissions. Differences in the estimates for fugitive dust can be attributed in part to the uncertainty of the fugitive dust inventory, which needs further improvement; the higher contribution from CMB might also mean that it includes the effect of dust from the local area and beyond.

Spatial Variations of Source Contributions

The sampling sites of BUU, CAPM, and CRAES in this study are far away from each other and in different parts of Beijing. Comparing source contributions at these sites could provide an outline of the spatial variations of $PM_{2.5}$ origins. As shown in Figure 4, comparable $PM_{2.5}$ mass concentrations and similar source contributions were observed at the three sampling locations; similar results were found in a study of São Paulo, a megacity in Brazil (Castanho and Artaxho, 2001). The similarity of $PM_{2.5}$ source contributions over a large area suggests that the air pollution caused by fine particles tends to be a regional problem and quite different from pollution caused by coarse particles.

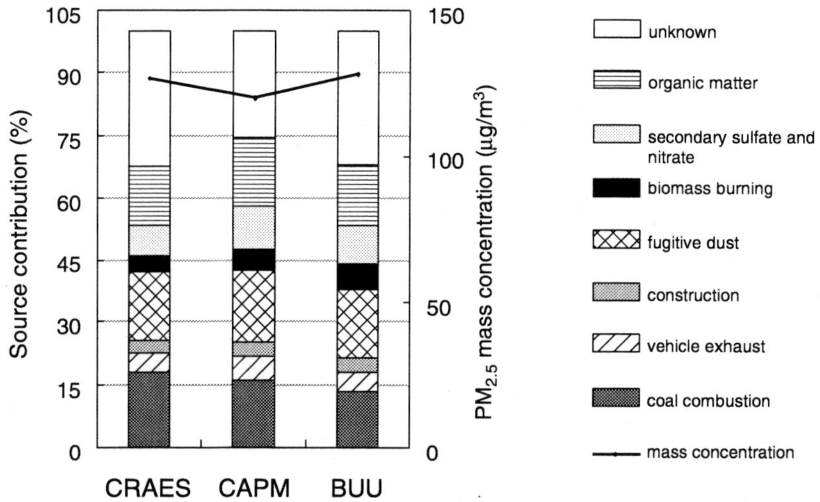

FIGURE 4 Source contributions at the three sampling sites in Beijing.

Seasonal Variations of Source Contributions

Figure 5 shows the seasonal pattern of contributions from major sources of $PM_{2.5}$. The contributions of coal combustion and fugitive dust change remarkably from season to season. The highest source contribution is in spring from fugitive dust and in winter from coal combustion. The dry weather in Beijing is the main reason for the severe pollution caused by fugitive dust in spring. The meteorological record for 2000 shows that the highest average wind velocity and lowest precipitation level occurred in the spring; days with little wind (less than 2 m/s velocity) or high humidity (higher than 60 percent) were rare (Wang and Zhang, 2002). In addition, dry weather in northern China since the 1990s and large areas of unpaved ground have increased dust pollution (Wang and Zhang, 2002). A sandstorm during the first two sampling days in spring can explain the very high contribution of dust.

In winter, the contribution of coal combustion was high both because of high consumption of coal and unfavorable meteorological conditions for dispersion. The results of a study by Zheng et al. (2002) showed similar conditions contributed to high contributions from wood combustion in cities in the southeastern United States. The results of that study showed that the contribution of wood combustion to the total OC concentration in the atmosphere increased during the colder months of October and January as a result of the high level of residential wood burning.

Unlike coal combustion and fugitive dust, vehicle exhaust and construction

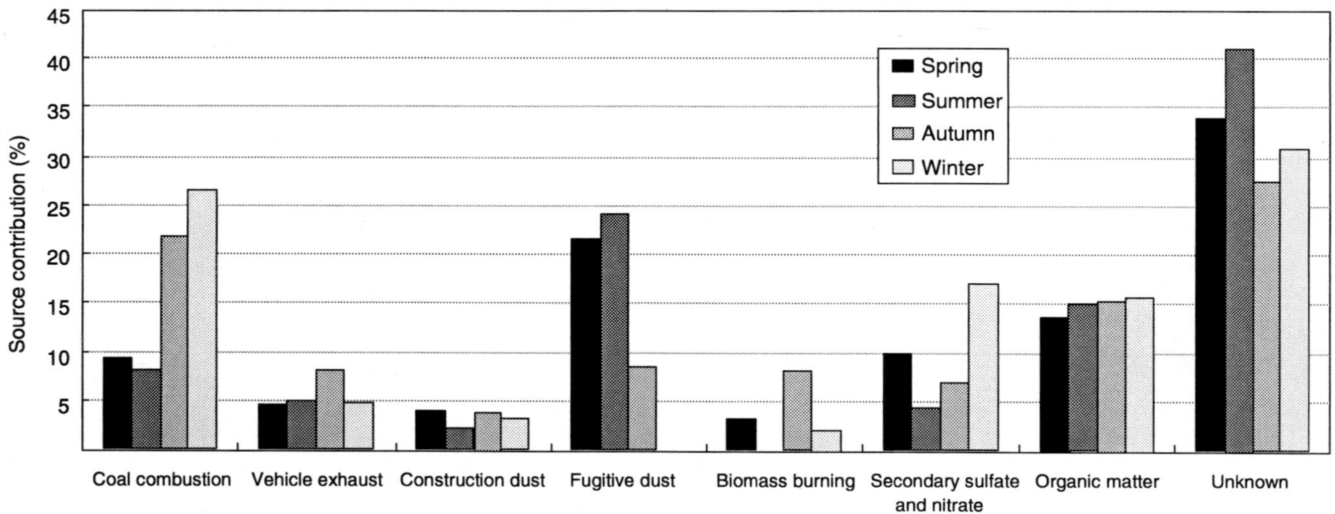

FIGURE 5 Seasonal variations in source contributions in Beijing.

dust contributions did not vary with the seasons. Emissions from biomass burning, however, varied a great deal with the seasons. The contribution of biomass burning was high in spring and autumn; the contribution in autumn increased by 8.1 percent, 2.5 times higher than in spring and 3.9 times higher than in winter. Using potassium as a tracer for biomass burning, Duan et al. (2001) investigated the seasonal variations and found that the concentration increased in Beijing in spring and autumn, which is consistent with the results in this study. The increases are attributable to crop-straw burning after the harvest in autumn and before planting in spring.

The contribution of secondary sulfate and nitrate, the result of the transformation of high-level SO_2 into secondary pollutants by homogeneous and heterogeneous reactions and the accumulation of pollutants in stagnant weather, rises to its highest level (17 percent) in winter, which is 1.7 times higher than in spring and 2.5 times higher than in autumn. However, because rainy or cloudy days dominated the sampling period in summer, the contribution of secondary sulfate and nitrate was not as high as expected.

The variability of the contribution of organic matter, from 13.7 to 15.8 percent, was not correlated with the seasons. By contrast, in the southeastern United States, the highest contribution of organic matter was in July and was closely correlated with concentrations of sulfate, nitrate, and ammonium (Zheng et al., 2002).

CONCLUSIONS

The study described in this paper explored the sources of $PM_{2.5}$ by developing a source inventory and applying the CMB model to ambient samples collected at three sites in Beijing at different times of the year. The primary sources responsible for most of the mass concentration of $PM_{2.5}$ are fugitive dust, coal combustion, and vehicle exhaust. Construction dust and biomass burning were also detected in the CMB model, but their contributions were relatively small. Moreover, about 9.6 percent of $PM_{2.5}$ mass concentration was attributed to secondary sulfate and nitrate, and 15 percent was attributed to organic matter. In addition, 6.0 to 10.6 percent of $PM_{2.5}$ could be explained by SOA. Thus, the contributions from secondary sources were as high as 15.6 to 20.2 percent, the result of the high oxidation capacity of the atmosphere and the rapid formation of secondary pollutants in the air.

Using the CMB model at the sampling sites provided some insight into the spatial variability of $PM_{2.5}$ sources. The results showed that the source contributions to $PM_{2.5}$ mass concentrations were similar at all three sampling locations. This might imply that fine particulate air pollution has extended beyond Beijing and is developing into a regional problem. In addition, distinct seasonal variations were observed for contributions of fugitive dust, coal combustion, biomass burning, and secondary sulfate and nitrate.

REFERENCES

Abbey, D.E., R.J. Burchette, S.F. Knutsen, W.F. McDonnell, M.D. Lebowitz, and P.L. Enright. 1998. Long-term particulate and other air pollutants and lung function in nonsmokers. American Journal of Respiratory Critical Care Medicine 158: 289–298.

Bencala, K.E., and J.H. Seinfeld. 1979. An air quality model performance assessment package. Atmospheric Environment 13: 1181–1185.

Bergin, M.H., G.R. Cass, J. Xu, C. Fang, L. Zeng, T. Yu, L.G. Salmon, C.S. Kiang, X.Y. Tang, Y.H. Zhang, and W.L. Chameides. 2001. Aerosol radiative, physical, and chemical properties in Beijing during June 1999. Journal of Geophysical Research 106(D16): 17969–17980.

Castanho, A.D.A., and P. Artaxo. 2001. Wintertime and summertime São Paulo aerosol source apportionment study. Atmospheric Environment 35: 4889–4902.

Duan, F., Y. Lu, Y. Di, X. Liu, H. Zhang, X. Yang, and T. Yu. 2001. Influence of straw burning on the air quality in Beijing. Environmental Monitoring in China 17(3): 8–11. In Chinese.

Gordon, G.E. 1980. Receptor models. Environmental Science and Technology 14: 792–800.

Harrison, R.M., D.J.T. Smith, C.A. Pio, and L.M. Castro. 1997. Comparative receptor modeling study of airborne particulate pollutants in Birmingham (United Kingdom), Coimbra (Portugal) and Lahore (Pakistan). Atmospheric Environment 31: 3309–3321.

Hildemann, L.M., G.R. Markowski, and G.R. Cass. 1991. Chemical composition of emissions from urban sources of fine organic aerosol. Environmental Science and Technology 25: 744–759.

Lin, J.J., and H. Tai. 2001. Concentrations and distributions of carbonaceous species in ambient particles in Kaohsiung City, Taiwan. Atmospheric Environment 35: 2627–2636.

Liu, M.K., and J.H. Seinfeld. 1975. On the validity of grid and trajectory models of urban air pollution. Atmospheric Environment 9: 555–574.

Ostro, B.D., S. Hurley, and M.J. Lipsett. 1999. Air pollution and daily mortality in the Coachella Valley, California: a study of PM_{10} dominated by coarse particles. Environmental Research 81: 231–238.

Schauer, J.J., and G.R. Cass. 2000. Source apportionment of wintertime gas-phase and particle-phase air pollutants using organic compounds as tracers. Environmental Science and Technology 34: 1821–1832.

Schauer, J.J., W.F. Rogge, L.M. Hildemann, M.A. Mazurek, and G.R. Cass. 1996. Source apportionment of airborne particulate matter using organic compounds as tracers. Atmospheric Environment 30: 3837–3855.

Schwartz, S.E. 1996. The whitehouse effect-shortwave radiative forcing of climate by anthropogenic aerosols: an overview. Journal of Aerosol Science 3: 359–382.

Song, Y., X. Tang, C. Fang, Y. Zhang, M. Hu, and L. Zeng. 2002a. Source apportionment on fine particles in Beijing. Environment Science 23(6): 11–16. In Chinese.

Song, Y., X. Tang, Y. Zhang, M. Hu, C. Fang, L. Zeng, and W. Wang. 2002b. Effect on fine particles by the continued high temperature weather in Beijing. Environment Science 23(4): 33–36. In Chinese.

Tang, X., J. Li, and D. Chen. 1995. Summertime photochemical pollution in Beijing. Pure Applied Chemistry 67: 1465–1468.

Wang, S., and X. Zhang. 2002. Meteorological features of PM_{10} pollution in Beijing. Journal of Applied Meteorological Science 13(Suppl.): 177–184. In Chinese.

Watson, J.G. 1984. Overview of receptor model principles. Journal of the Air Pollution Control Association 34(6): 619–623.

Watson, J.G., J.C. Chow, and J.E. Houck. 2001. $PM_{2.5}$ chemical source profiles for vehicle exhaust, vegetative burning, geological material, and coal burning in northwestern Colorado during 1995. Chemosphere 43(8): 1141–1151.

Wilson, W.E., and H.H. Suh. 1997. Fine particles and course particles: concentration relationships relevant to epidemiological studies. Journal of the Air and Waste Management Association 47: 1238–1249.

Yantin, M., S. Tuncel, N.K. Aras, I. Olmez, S. Aygun, and G. Tuncel. 2000. Atmospheric trace elements in Ankara, Turkey: 1. Factors affecting chemical composition of fine particles. Atmospheric Environment 34: 1305–1318.

Zhang, Y., M. Shao, M. Hu, and X. Tang. 1997. Air Quality in Beijing and Its Transition from Coal Burning Caused Problems to Traffic Exhaust Related Pollution. In Proceedings of the Air and Waste Management Association 90th Annual Meeting and Exposition. Pittsburgh, Pa.: Air and Waste Management Association. Available on CD-ROM AM 97-CD.

Zhang, Y., K. Shao, X. Tang, and J. Li. 1998. The study of urban photochemical smog pollution in China. Acta Scientiarum Naturalium Universitatis Pekinensis 34(2-3): 392–400. In Chinese.

Zheng, M., G.R. Cass, J.J. Schauer, and E.S. Edgerton. 2002. Source apportionment of $PM_{2.5}$ in the southeastern United States using solvent-extractable organic compounds as tracers. Environmental Science and Technology 36: 2361–2371.

Radiative Forcing by Anthropogenic Aerosols: Sources and Impacts

MICHAEL H. BERGIN
Georgia Institute of Technology

China is the most populous and fastest growing nation in the world. In addition, China's gross domestic product is growing by ~10 percent per year (Freemantle, 1998). This extraordinary economic growth is associated with increases in anthropogenic pollutants, such as aerosols and ozone precursors, from industrial sources that will increase dramatically in the future. Sulfur emissions (from the burning of coal, which currently supplies ~75 percent of China's total energy) are expected to double by 2010 (Daniel, 1994; Wolf and Hidy, 1997). The impact of anthropogenic pollutants, particularly aerosols, on the environment in this part of the world remains uncertain, however, because we have only a few measurements of the chemical, physical, and radiative properties of aerosols on relevant temporal and spatial scales.

A recent study of temperature trends in China over the past ~40 years shows that the temperature over a relatively large area of the country has increased by ~0.2–0.4°C, which is attributable to a concomitant increase in greenhouse gases (Li et al., 1995). At the same time, the temperature over a large part of southern China, extending from Sichuan Province to the Yangtze delta region, dropped by ~0.2–0.4°C. The cooling trend has been evident since the mid-1970s, which coincides with the start of economic reforms and intense industrialization in the area and has been accompanied by significant decreases in visual range, attributed to increases in anthropogenic aerosol loading. Li et al. (1995) suggest that the cooling is due to the scattering of shortwave radiation out of the Earth's atmosphere during clear-sky conditions; this is known as "direct," shortwave, aerosol, radiative forcing.

These results are consistent with the model estimates of Chameides et al.

(1999) suggesting that anthropogenic sulfur and carbonaceous aerosols may reduce the surface irradiance in these regions by as much as 30 percent, which may result in a decrease in crop yield. Modeling results based on projections of future sulfur emissions also predict cooling (Engardt and Rodhe, 1998; Wolf and Hidy, 1997).

China is also a significant source of black carbon (also called elemental carbon) aerosols from incomplete combustion processes. Black carbon absorbs solar radiation, which results in warming of the atmosphere. Using a global climate model to study the influence of black carbon emissions in China on both regional and global climate, Menon et al. (2002) found that increases in black carbon emissions correlated roughly with observed severe weather patterns (e.g., heavy rainfall and droughts) and influenced regional temperatures as far away as North America.

"Indirect" forcing of climate by aerosols emitted in China may also affect global climate. The indirect effects are attributable to changes in the optical properties of clouds associated with the presence of anthropogenic aerosol particles. Although much of China has a characteristic haziness, not much is known about the sources of aerosols and their impact on radiation balance, climate, and plant growth.

CONCENTRATIONS AND SOURCES OF AEROSOLS IN CHINA

Although little information is available on aerosol radiative properties in China, several studies report extremely high mass concentrations of aerosols in urban areas. Measurements of total suspended-particulate (TSP) mass made over a one-year period in several major cities averaged ~300 μg/m^3; major sources include windblown dust and fossil fuel combustion (Hashimoto et al., 1994). This value is roughly four times the maximum allowable in the United States, according to the national ambient air quality standard (NAAQS) for TSP, which is 75 μg/m^3. Waldman et al. (1991) report values of aerosol particle mass with diameters of less than 2.5 μm ($PM_{2.5}$) over a two-week intensive sampling period of 139 μg/m^3 (with a range of 54 to 207 μg/m^3) in Wuhan, an industrialized city in central China. Particles of this size may not only accumulate in the lungs, but may also scatter and absorb light in visible wavelengths; the mean value is greater by more than a factor of 2 than the proposed U.S. $PM_{2.5}$ NAAQS 24-hour average value of 50 μg/m^3 (Waggoner and Weiss, 1980). Based on a study by Waldman et al. (1991), the contributions of sulfate, nitrate, and carbonaceous aerosol to $PM_{2.5}$ are ~30 percent, 20 percent, and 50 percent, respectively. A recent year-long monitoring study in Beijing also found relatively high values of $PM_{2.5}$, with a reported annual mean concentration of ~120 μg/m^3 (Cao et al., 2003). The results showed that organic carbon, sulfate, and nitrate collectively account for ~60 percent of the fine particle mass in Beijing. Measurements made at several stations in the Pearl River delta region during January and February of

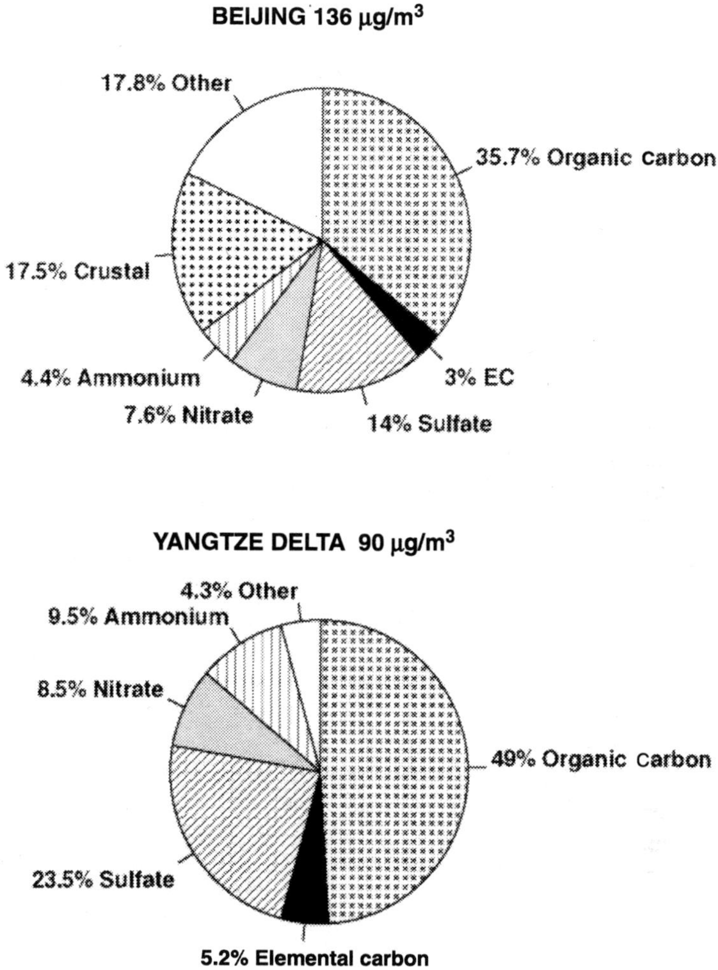

FIGURE 1 The $PM_{2.5}$ chemical composition in Beijing (June 1999) and at a location in the Yangtze Delta (November 1999). Sources: Bergin et al., 2001; Xu et al., 2002.

2001 found a mean $PM_{2.5}$ concentration of 70 µg/m^3, with nearly 40 percent of the fine particle mass composed of organic compounds (Cao et al., 2003).

Figure 1 shows the chemical composition of aerosols impacted by different sources in two locations, Beijing in June 1999 (Bergin et al., 2001) and Linan (in the Yangtze delta approximately 200 km southwest of Shanghai) in November 1999 (Xu et al., 2002). At both locations, the $PM_{2.5}$ concentrations exceeded both the daily and annual mean U.S. NAAQS values. The chemical composition of fine particulate matter was dominated by carbonaceous aerosol and sulfate. The organic carbon has a variety of sources, including diesel and automobile exhaust, the burning of biomass, cooking, the combustion of coal, and biogenic emissions from plants (Schauer et al., 1996). Mobile sources, as well as the burning of coal and biomass, are the main contributors to the organic carbon concentrations in Beijing. In addition, the effects of coal burning are reflected in the relatively large contribution of sulfate (14 percent) to the $PM_{2.5}$. A characteristic of the Beijing aerosol is the presence of crustal species that come from windblown dust. Arid regions of China, particularly the Gobi and Takla Makan deserts, are upwind sources of dust for much of northeastern China (Zhang et al., 2002). Dust storms are particularly intense in the spring; dust from these storms has been detected as far away as Greenland (Bory et al., 2003) and North America (McKendry et al., 2001).

A surprising finding is the relatively high $PM_{2.5}$ concentrations in the Yangtze delta region, because the sampling site is located in an agricultural area that does not have a great deal of industrial activity. Despite this, the mean $PM_{2.5}$ concentration is 90 $\mu g/m^3$, which is as high as concentrations measured in many urban areas of China. The most revealing difference between the chemical compositions in Linan and Beijing is the dominance of particulate organic carbon, which accounts for nearly half of the aerosol mass in the Yangtze delta. Elemental carbon, a product of incomplete combustion, is also comparatively high in the Yangtze Delta.

Although the sources of carbonaceous aerosol are not definitively known, the measurements in the Yangtze delta were made during the fall rice harvest, when the burning of crop residue is widespread throughout the region. Therefore, it would appear that the carbonaceous component of the aerosol in the Yangtze delta is primarily linked with the burning of biomass. In Beijing, particulate organic compounds are more likely from many different sources.

Although dust from the arid regions of China has been known to reach the Yangtze delta and more southerly regions of China, the contribution of dust to fine particulate mass is generally negligible in the measurements shown in Figure 1. The relatively high sulfate concentrations in the Yangtze delta, which account for ~24 percent of the $PM_{2.5}$ mass, are probably from a combination of regional sources and the long-range transport of air masses carrying particulates from regions to the west, such as Sichuan Province. Thus, the rural region of the Yangtze delta, one of the most important agricultural areas in China, appears to have relatively poor air quality as a result of both local emissions and the transport of polluted air from industrialized regions.

Overall, many of the studies suggest extremely high aerosol loadings in urban and some rural areas of China, with primary anthropogenic contributions to fine particulate mass of sulfate and carbonaceous compounds. Streets et al. (2003), who have compiled emissions inventories in Asia for gaseous and primary aerosol precursors, found that anthropogenic sulfur dioxide (SO_2) emissions, which are responsible for most of the sulfate in particulate matter in

China, come primarily from coal burning for power generation (50 percent) and industrial purposes (36 percent); a smaller, but still significant contribution comes from the domestic use of coal. A fraction of industrial SO_2 emissions are linked to the generation of electricity by industry. The emission inventories estimated by Streets et al. (2003) are within several percent of independent estimates made by the Chinese government (SEPA, 2000).

Streets et al. also estimate that black carbon in China is dominated by the residential burning of coal and biofuels, which together account for ~75 percent of all emissions; in urban areas, however, black carbon emissions may be mostly from mobile sources (i.e., diesel and gasoline vehicles). Emissions of organic carbon are currently poorly understood. Streets et al. (2003) estimate the sources to be dominated by residential fuel combustion (~76 percent) and the burning of biomass (22 percent). The burning of coal for power generation cannot be ruled out, however, as a significant source of organic carbon, particularly in urban regions.

One must be very cautious about inferring the sources of carbonaceous aerosols based on current emissions inventories. Streets et al. set an uncertainty level of ±450 percent for organic carbon emissions. Data from two recent field experiments, the Asian Pacific Regional Aerosol Characterization Experiment (ACE-ASIA) and Transport and Chemical Evolution Over the Pacific (TRACE-P), that focus partly on concentrations of particulate matter and related precursors in East Asia, will soon be available. The results of these two studies should shed additional light on the chemical composition, as well as the effects, of aerosols on the radiation balance in the vicinity of China.

RADIATIVE FORCING BY AEROSOLS

Although several studies report the chemical and physical properties of aerosols, very few measurements on pertinent temporal and spatial scales are available on the radiative properties that govern the scattering and absorption of solar radiation by aerosol particles. As a result, estimates of the influence of aerosols on climate must be based on model estimates of aerosol loading and properties. Figure 2 (based on a model by Chameides et al., 1999) shows the estimated annual mean aerosol optical depth at a wavelength of 550 nanometers (i.e., the vertical profile of the aerosol light-extinction coefficient) based on SO_2 emissions and the assumption that organic carbon and sulfate contribute similarly to fine particulate mass throughout China. Although the model estimates probably reflect the impact of SO_2 emissions on aerosol loadings, they only very roughly reflect the influence of particulate organic carbon, because neither elemental carbon nor organic carbon emissions are taken into account directly. Nor does the model take into account the contribution of dust to annual aerosol loading, which, as discussed above, can account for a significant fraction of the fine particulate mass, particularly during the spring season in much of northern China. Even though the results in Figure 2 are only rough estimates of the annual aerosol optical depth throughout the model domain, they agree qualitatively with actual measurements, which are also shown.

Relatively high aerosol loadings can be seen throughout China, with a "hot spot" centered in the Sichuan region and extending east over the Yangtze delta. The model results clearly indicate that upwind sources influence aerosol loadings in the Yangtze delta. Based on these data, we can also estimate the reduction in surface irradiance due to light extinction by aerosols (Figure 3). Surface irradiance is reduced annually by nearly 30 percent in the vicinity of the Sichuan region, with values of 10 to 15 percent throughout much of eastern China. Xu et al. (2003) combined these data with atmospheric radiative transfer modeling, which shows an annual mean top-of-the-atmosphere, direct aerosol radiative forcing of approximately –12 W/m^2 in the Yangtze delta. This value is nearly an order of magnitude greater than the global mean value suggested by the Intergovernmental Panel on Climate Change (IPCC, 2001). Thus, direct scattering and absorption by aerosols can have a dramatic effect on radiation balances at the top of the atmosphere, as well as at the surface of the Earth.

Given these extremely high aerosol loadings, aerosols probably also have an

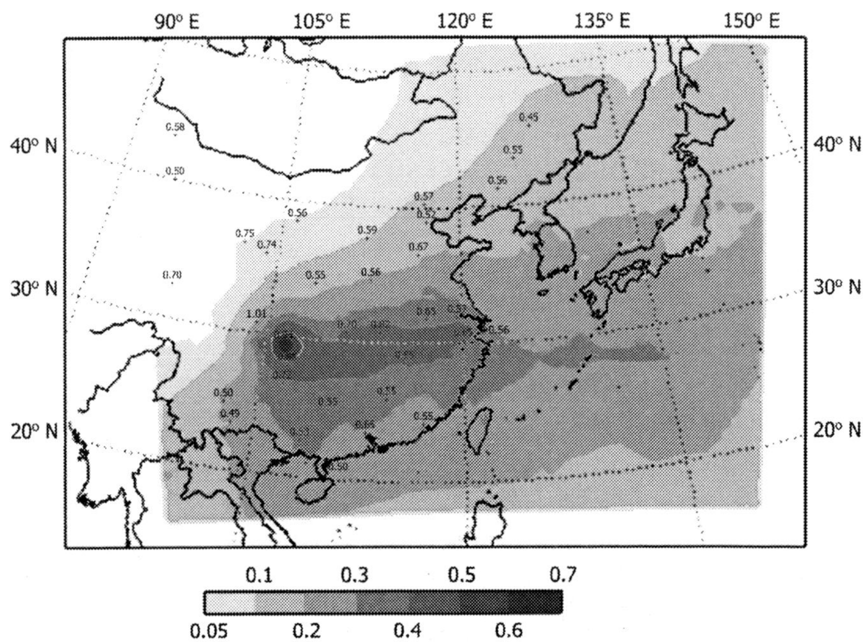

FIGURE 2 Annual average aerosol optical depth at 550 nm (τ_a) estimated by the model of Chameides et al., 1999. Values indicate measurements.

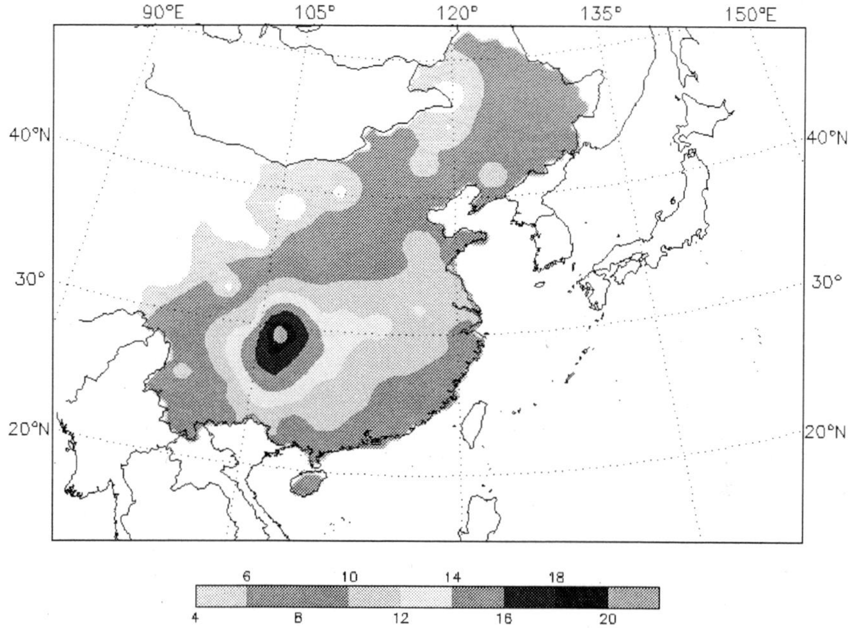

FIGURE 3 Estimated annual percentage change in surface irradiance due to the presence of aerosols. Source: Chameides et al., 1999.

indirect effect on the radiation balance. The indirect effect, which results from the modification by aerosols of cloud albedo, amount, and lifetime, is extremely difficult to quantify because of our current lack of understanding of the processes that govern aerosol-cloud interactions. In fact, the uncertainty in estimates of indirect aerosol effects (from 0 to -2 W/m^2) dominates the overall uncertainty in estimates of anthropogenic climate forcing (IPCC, 2001).

When Chameides et al. (2002) compared model estimates of aerosol loading over China and East Asia with satellite-derived cloud properties, including optical depth and amount, they found a surprisingly strong correlation ($r^2 \sim 0.6$). In other words, regions of East Asia with high aerosol loadings also generally have more reflective clouds. Based on these findings, the indirect radiative forcing is estimated to be approximately 1.5 times the direct radiative forcing. This would mean, for instance, that, based on the direct estimates of Xu et al. (2002) (discussed above), the indirect aerosol radiative forcing in the Yangtze delta region is roughly -18 W/m^2. Thus, the combined aerosol radiative forcing over much of China impacted by aerosols, taken as the sum of the direct and indirect effects, may be from -30 to -40 W/m^2. These values are much more than an order of magnitude greater than the overall annual mean radiative forcing (including greenhouse gases and aerosols) of a few W/m^2 estimated by IPCC. Thus, we can conclude that the emissions of aerosols and their precursors in China are having a significant influence on the radiation balance of Earth and very likely significant climatic impacts in East Asia and perhaps globally (Menon et al., 2003).

INFLUENCE OF AEROSOLS ON CROPS

Because aerosols influence the balance of surface radiation, they may also influence plant growth by changing the amount of photosynthetically available radiation (PAR, i.e., radiation between 400 nm and 700 nm). Figure 4 shows the change in net PAR (NPAR) reaching the surface (NPAR is the difference between downward and upward PAR) as a function of aerosol optical depth at 500 nm (τ_{500}) measured in the Yangtze delta region during November 1999. NPAR and aerosol optical depth are both normalized to account for the change in

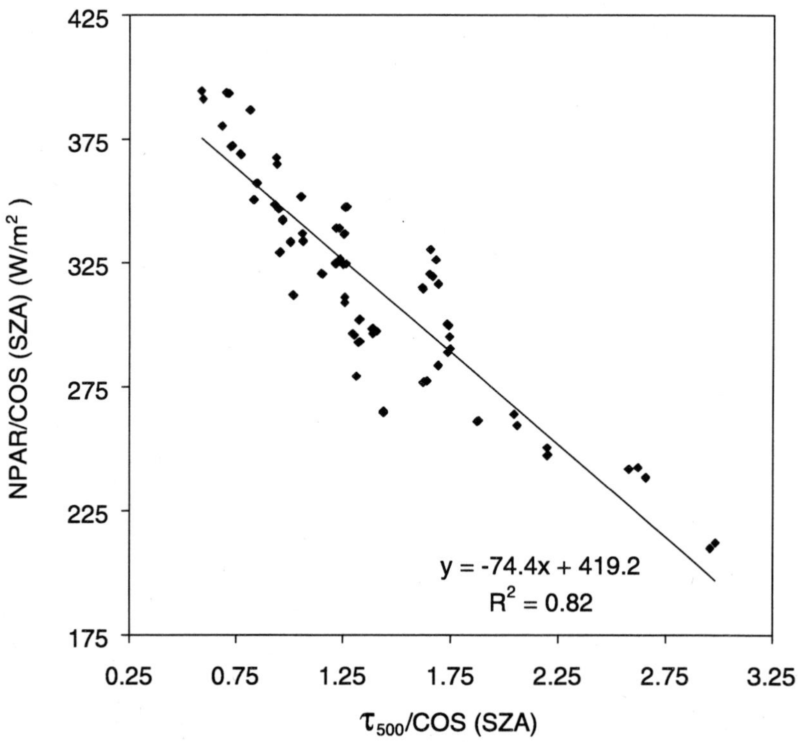

FIGURE 4 Change in net PAR (NPAR) reaching the surface in the Yangtze delta as a function of aerosol optical depth for clear-sky conditions. (Both are normalized to account for the change in solar zenith angle [SZA].) Source: Xu et al., 2003.

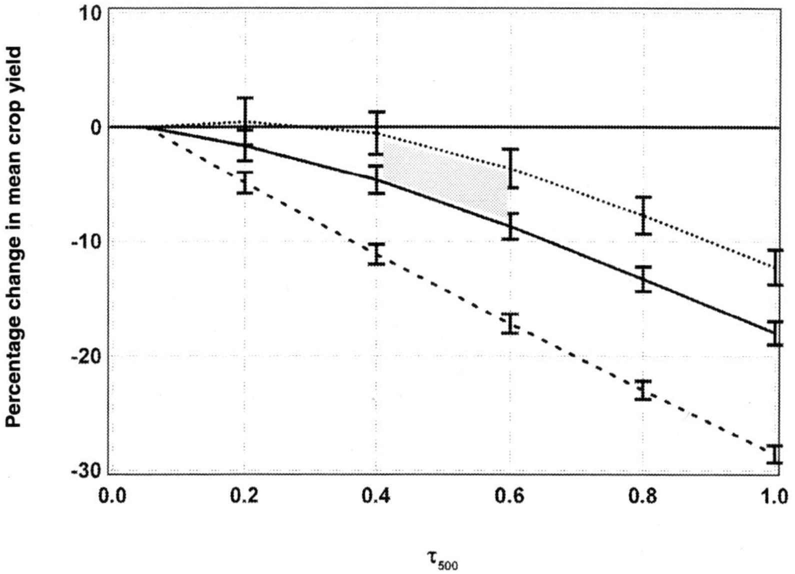

FIGURE 5 Estimated percentage change in rice yield as a function of aerosol optical depth for radiation use efficiency increases of 0 percent (dashed line), 50 percent (solid line), and 100 percent (gray line). Source: Greenwald et al., 2003.

solar zenith angle (SZA). Figure 4 shows that the amount of PAR reaching the surface decreases as aerosol loading increases. The slope of the line, commonly called the direct aerosol radiative forcing efficiency, indicates a decrease in NPAR of ~74 W/m^2 for each unit of aerosol optical depth. Based on Figure 4 and measurements of aerosol properties, Xu et al. (2003) estimate that aerosols decrease the amount of PAR reaching the surface by ~15 to 20 percent; this is relatively consistent with the model estimates presented in Figure 3.

The attenuation of solar radiation by aerosols results in a decrease in the amount of direct solar radiation that reaches the surface, and, at the same time, an increase in diffuse radiation. In a plant canopy, an increase in atmospheric aerosol loading results in a decrease in PAR illuminating leaves that are normally sunlit and an increase in PAR for leaves that are shaded. Thus, as the aerosol optical depth increases, the relative amount of diffuse (and total) radiation reaching a plant canopy generally increases. Therefore, a decrease in surface PAR does not necessarily result in a proportionate decrease in plant growth. This is illustrated in Figure 5, which shows estimates of the change in rice yield as a function of aerosol loading for meteorological conditions similar to those of the Yangtze delta.

The estimates in Figure 5 are based on a coupled atmospheric radiative transfer-crop model described by Greenwald et al. (2003). The lines represent three separate scenarios of radiation use efficiency (RUE, i.e., the mass of carbon fixed by a plant per MJ PAR absorbed) based on the relative amount of diffuse-to-total PAR reaching the plant canopy. The bottom (dashed) line represents no increase in RUE as a function of aerosol optical depth (and increasing diffuse-to-total radiation ratio). The top line assumes a maximum increase in RUE of 100 percent with increasing diffuse radiation. The gray area shown in the middle is the most likely expected crop yield, based on limited knowledge of rice crop response as a function of diffuse fraction and aerosol optical depth values in China. The estimates suggest that the decrease in crop yield ranges from a few percent to nearly 10 percent. These are rough estimates, of course, because of uncertainties in the aerosol loadings, crop response to changes in radiation, nutrient and water stresses, and meteorology. Nevertheless, the results suggest that a decrease in the sources of aerosols may lead to an increase in crop yield.

Another possible influence on plant growth is the deposition of particulate matter on leaves. In addition to damage to the leaves from acidity, aerosol deposits can scatter and absorb radiation, thus reducing PAR for photosynthesis. Particles that are insoluble in water pose the greatest threat because they cannot be easily washed from leaves by precipitation; thus, they accumulate over time. Measurements in the Yangtze delta suggest that a significant fraction of fine particulate matter (~30 to 40 percent) is not soluble in water. Given the relatively large concentration of fine particulate matter observed at many locations in China, the attenuation of PAR by deposits of particles may significantly decrease available PAR and hence plant photosynthesis. The deposition of dust particles, which are primarily insoluble in water, also decreases PAR in regions of China that experience high dust loading.

Figure 6 shows estimates of the decrease in PAR available for photosynthesis caused by dry deposition (EX_{PAR}), based on measurements in the Yangtze delta of elemental carbon and water-insoluble aerosol mass concentration (Bergin et al., 2001). The combined scattering and absorption of deposited aerosols can account for a nearly 30 percent decrease in the amount of available PAR reaching leaves over a two-month period. The attenuation of PAR at the leaf surface is the result of absorption by elemental carbon and scattering by water-insoluble carbonaceous aerosols. Thus, in addition to attenuation of PAR by atmospheric aerosols, particles deposited on leaves may also influence photosynthesis and plant growth.

SUMMARY

Atmospheric particulate matter in China is very likely having a wide range of impacts, including damage to human health, reduced visibility, modified climate, and decreased plant growth. Based on recent measurements, the main contributors to fine particulate matter are sulfate and organic compounds. Sulfate originates

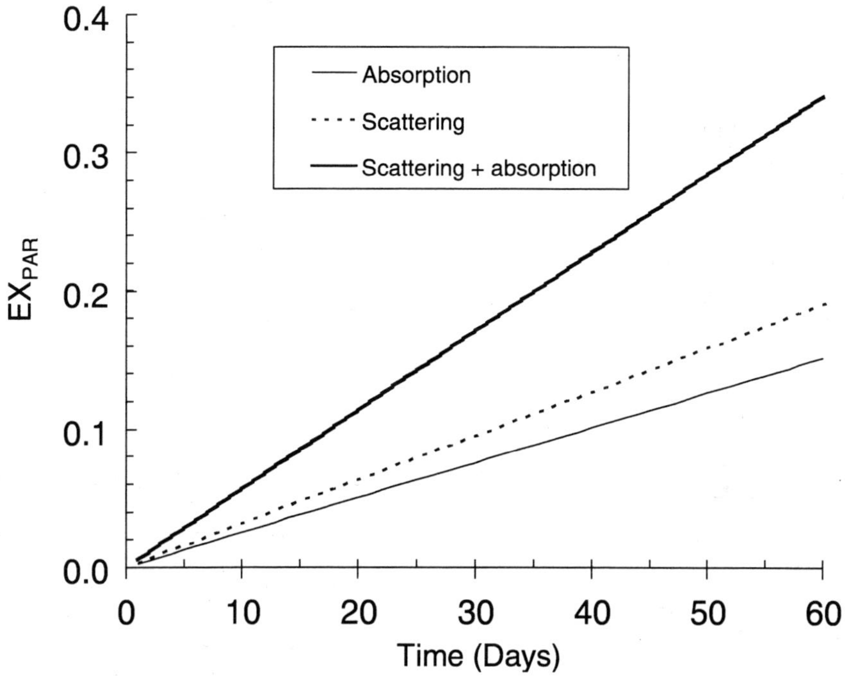

FIGURE 6 Estimated change in PAR available for plant photosynthesis, EX_{PAR}, due to deposition of elemental carbon and water-insoluble aerosol particles estimated for the Yangtze delta as a function of time. Source: Bergin et al., 2001.

primarily from the burning of coal for electricity and industrial purposes, with lesser contributions from residential sources. Based on measurements in both urban and rural locations, organic compounds appear to dominate the fine particulate mass. The sources of these compounds are currently not well understood but probably include diesel and automobile emissions, the burning of coal and biomass, and cooking. Clearly, future research must address the sources of these compounds. In relatively arid regions, dust can also be a significant fraction of the aerosol loading.

Anthropogenic aerosols have been found to have a profound influence on the local, regional, and even global radiation balance. Model estimates show that aerosols may be decreasing the radiation reaching the surface by as much as 30 percent in many parts of China. This decrease is associated with the scattering and absorption of light by anthropogenic aerosols. Recent model estimates suggest that the aerosol loadings over China may be influencing weather not only in China, but also elsewhere around the globe. In rough agreement with model estimates, measurements in an agricultural region of the Yangtze delta indicate a 15 to 20 percent reduction in the amount of PAR reaching the surface, which may reduce crop production by as much as 10 percent in the region. The relatively high concentrations of fine particulate matter in the Yangtze delta may also result in the deposition of particulate matter on leaves, which may also affect plant growth by scattering and absorbing radiation. Overall, aerosols appear to have a negative influence on crop growth, and perhaps plant growth, in many regions of China.

REFERENCES

Bergin, M.H., R. Greenwald, J. Xu, Y. Berta, and W.L. Chameides. 2001. Influence of aerosol dry deposition on photosynthetically active radiation available to plants: a case study in the Yangtze delta region of China. Geophysical Research Letters 28(18): 3605–3608.

Bory, A.J.M., P.E. Biscaye, and F.E. Grousset. 2003. Two distinct seasonal Asian source regions for mineral dust deposited in Greenland (NorthGRIP). Geophysical Research Letters 30(4): 1167.

Cao, J.J., S.C. Lee, K.F. Ho, X.Y. Zhang, S.C. Zou, K. Fung, J.C. Chow, and J.G. Watson. 2003. Characteristics of carbonaceous aerosol in Pearl River delta region, China, during 2001 winter period. Atmospheric Environment 37(11): 1451–1460.

Chameides, W.L., C. Luo, R. Saylor, D. Streets, Y. Huang, M. Bergin, and F. Giorgi. 2002. Correlation between model-calculated anthropogenic aerosols and satellite-derived cloud optical depths: indication of indirect effect? Journal of Geophysical Research-Atmospheres 107(D10): 4085.

Chameides, W.L., H. Yu, S.C. Liu, M. Bergin, X. Zhou, L. Mearns, G. Wang, C.S. Kiang, R.D. Saylor, C. Luo, Y. Huang, A. Steiner, and F. Giorgi. 1999. Case study of the effects of atmospheric aerosols and regional haze on agriculture: an opportunity to enhance crop yields in China through emission controls? Proceedings of the National Academy of Sciences 96(24): 13626–13633.

Daniel, M. 1994. Chinese coal prospects to 2010. Paper #11. Paris: International Energy Agency.

Engardt, M., and H. Rodhe. 1993. A comparison between patterns of temperature trends and sulfate aerosol pollution. Geophysical Research Letters 20(2): 117–120.

Freemantle, M. 1998. A makeover for science in China. Chemical and Engineering News 76(34): 17–27.

Greenwald, R., M.H. Bergin, J. Xu, D. Cohan, G. Hoogenboom, and W.L. Chameides. 2003. The influence of aerosols on crop production: a study using the CERES crop model. Journal of Geophysical Research-Atmospheres, in review.

Hashimoto, Y., Y. Sekine, H.K. Kim, Z.L. Chen, and Z.M. Yang. 1994. Atmospheric fingerprints of East Asia, 1986–1991: an urgent record of aerosol analysis by the Jack Network. Atmospheric Environment 28(8): 1437–1445.

IPCC (Intergovernmental Panel on Climate Change). 2001. Climate Change 2001: The Scientific Basis. Cambridge, U.K.: Cambridge University Press.

Li, X., X. Zhou, W. Li, and L. Chen. 1995. The cooling of Sichuan Province in the recent 40 years and its probable mechanism. Acta Meteorologica Sinica 9: 57–68.

McKendry, I.G., J.P. Hacker, R. Stull, S. Sakiyama, D. Mignacca, and K. Reid. 2001. Long-range transport of Asian dust to the Lower Fraser Valley, British Columbia, Canada. Journal of Geophysical Research-Atmospheres 106(D16): 18361–18370.

Menon, S., J. Hansen, L. Nazarenko, and Y.F. Luo. 2003. Climate effects of black carbon aerosols in China and India. Science 297(5590): 2250–2253.

Schauer, J.J., W.F. Rogge, L.M. Hildemann, M.A. Mazurek, and G.R. Cass. 1996. Source apportionment of airborne particulate matter using organic compounds as tracers. Atmospheric Environment 30(22): 3837–3855.

SEPA (State Environmental Protection Administration). 2000. Report on the State of the Environment in China. Beijing: State Environmental Protection Administration of China.

Streets, D.G., T.C. Bond, G.R. Carmichael, S.D. Fernandes, Q. Fu, D. He, Z Klimont, S.M. Nelson, N.Y. Tsai, M.Q. Wang, J.-H. Woo, and K.F. Yarber. 2003. An inventory of gaseous and primary aerosol emissions in Asia in the year 2000. Journal of Geophysical Research-Atmospheres 108(D21): 8809.

Waggoner, A.P., and R.E. Weiss. 1980. Comparison of fine particle mass concentration and light-scattering extinction in ambient aerosol. Atmospheric Environment 14(5): 623–626.

Waldman, J.M., P.J. Lioy, M. Zelenka, L. Jing, Y.N. Lin, Q.C. He, Z.M. Qian, R. Chapman, and W.E. Wilson. 1991. Wintertime measurements of aerosol acidity and trace-elements in Wuhan, a city in central China. Atmospheric Environment Part B-Urban Atmosphere 25(1): 113–120.

Wolf, M.E., and G.M. Hidy. 1997. Aerosols and climate: anthropogenic emissions and trends for 50 years. Journal of Geophysical Research-Atmospheres 102(D10): 11113–11121.

Xu, J., M.H. Bergin, X. Yu, G. Liu, J. Zhao, C.M. Carrico, and K. Baumann. 2002. Measurement of aerosol chemical, physical and radiative properties in the Yangtze delta region of China. Atmospheric Environment 36(2): 161–173.

Xu, J., M.H. Bergin, R. Greenwald, and P.B. Russell. 2003. Direct aerosol radiative forcing in the Yangtze delta region of China: observation and model estimation. Journal of Geophysical Research-Atmospheres 108(D2): 4060.

Zhang, J., Y. Wu, C.L. Liu, Z.B. Shen, and Y. Zhang. 2002. Major components of aerosols in north China: desert region and the Yellow Sea in the spring and summer of 1995 and 1996. Journal of the Atmospheric Sciences 59(9): 1515–1532.

The Power Sector

Analysis of Emissions, Exposures, and Risks of Toxic Air Emissions from U.S. Coal-Fired and Oil-Fired Power Plants

CHRIS G. WHIPPLE
ENVIRON International Corporation

Coal has been mined in the United States since the mid-1700s and used for the production of electric power since about 1890. Major national efforts to control the environmental effects of coal combustion began in the last half of the twentieth century, but local efforts began much earlier. The cities of Chicago and Cincinnati enacted clean air legislation in 1881. In 1904, Philadelphia passed an ordinance "to regulate emission of smoke from chimneys, stacks, flues or open spaces, providing a color scale for the measurement of the degree of darkness of such smoke, making it unlawful to permit the escape of smoke of a certain degree of darkness and providing a penalty for the violation of the ordinance. The Bureau of Boiler Inspectors enforced it" (Philadelphia Department of Public Health, 2004).

In 1955, Congress passed the Air Pollution Control Act. The first Clean Air Act was passed in 1963, and amendments were enacted in 1965, 1967, and 1969. In 1970, the Environmental Protection Agency (EPA) was formed, and significant changes were made to the Clean Air Act. Previous versions had provided mainly for research and development and left the regulation of air quality to the states, but the 1970 Act required that national primary and secondary ambient air quality standards be set. Primary standards applied to the protection of health, secondary standards to adverse economic impacts. In 1971, EPA set health standards for emissions of sulfur dioxide (SO_2). In 1977, President Carter's National Energy Plan recommended that all new power plants install best available pollution control technology. At that time, scrubbing was the only efficient process for removing high SO_2. In the following decade, EPA established national ambient standards for particulates, carbon monoxide, nitrogen oxides (NO_x), and ozone.

In 1980, Congress enacted the National Acid Precipitation Assessment Program (NAPAP) Study, a $555-million, 10-year research program for the study of "acid rain." Industries spent more than $1 billion on air pollution control equipment that year.

Prior to the passage of the 1990 Clean Air Act Amendments, emissions of trace substance "air toxics" from power plants were not regulated. EPA did have the authority, however, to regulate radionuclide emissions from power plants, and in the 1980s EPA established national emission standards for hazardous air pollutants (NESHAPS) for several sources of radionuclide emissions to air. In 1979 and 1984, EPA considered including coal- and oil-fired power plants as regulated sources of radionuclide emissions but ultimately determined that this was not necessary to protect public health.

ANALYSES BY THE ENVIRONMENTAL PROTECTION AGENCY AND THE ELECTRIC POWER RESEARCH INSTITUTE

In Section 112(n)(1)(A) of the 1990 Clean Air Act Amendments, Congress directed EPA to "perform a study of the hazards to public health reasonably anticipated to occur as a result of emissions by electric utility steam-generating units of . . . hazardous air pollutants." The amendments define an "electric utility steam-generating unit" as "any fossil-fuel-fired combustion unit of more than 25 megawatts electric (MWe) that serves a generator that produces electricity for sale." The amendments also required that "EPA proceed with rulemaking activities under section 112 to control HAP [hazardous air pollutant] emissions from utilities if EPA finds such regulation is appropriate and necessary after considering the results of the study."

The 1990 Clean Air Act Amendments also directed EPA to prepare "a study of mercury emissions from electric utility steam generating units, municipal waste combustion units, and other sources, including area sources. Such study shall consider the rate and mass of such emissions, the health and environmental effects of such emissions, technologies which are available to control such emissions, and the costs of such technologies."

The EPA mercury study, written in response to the 1990 amendments, was published in December 1997 (EPA, 1997). The EPA electric utility air toxics study, published in February 1998, includes an air toxics inhalation risk assessment for every fossil-fired steam-electric power plant in the United States with a capacity of 25 MWe or more.

In parallel with the EPA analyses, the U.S. electric utilities funded their own analyses of power plant air toxics by the Electric Power Research Institute (EPRI). Like EPA, the EPRI-sponsored studies included characterizations of emissions under various power-generation and emission-control scenarios, estimates of the cost and effectiveness of various control technologies, an assessment of air dispersion, an assessment of exposure through inhalation for all U.S. plants and exposure through multiple pathways (e.g., atmospheric deposition on food and through water) for case-study plants, and an assessment of the health risks associated with the estimated exposures (EPRI, 1994). (EPRI continued to sponsor similar studies after 1994.)

The results of the analyses are described in the remainder of this paper, based on the major elements of the analyses by EPA and EPRI listed below:

- scenarios for future power-plant inventories and use, including fuel mix, coal sources, and emission controls
- identification of substances included in the study
- emission factors for each fuel type and plant configuration
- air dispersion modeling
- exposure modeling for various exposure scenarios
  - calculations of inhalation exposures for all U.S. plants
  - calculations of multipathway exposures for selected chemicals for model plants (EPA) or case-study plants (EPRI)
- assessment of health effects from carcinogens and noncarcinogens
- assessment of inhalation risks
- assessment of multipathway inhalation risks
- characterization of results, including uncertainties

Both EPA and EPRI calculated inhalation exposures and risks for the more than 600 U.S. fossil-fuel, steam-electric power plants with a capacity of 25 MWe or more. Exposures and risks by the inhalation pathway were of particular interest for substances not considered likely to accumulate in the food chain.

However, another approach had to be used for some substances, such as mercury, for which the food pathway is more important than the inhalation pathway. EPA addressed exposures and risks from pathways other than inhalation by analyzing several model plants configured to be representative of common plant configurations. EPRI performed similar analyses using actual power plants, rather than hypothetical model plants. The primary focus of these multipathway analyses was on arsenic, mercury, and organics, such as dioxins and furans.

For radionuclides, EPA's office of radiation conducted a separate study of all U.S. plants using CAP88-PC, the model EPA uses for compliance analyses for air emissions of radioactive materials. EPRI did not conduct analyses for all U.S. plants but did analyze several case-study plants. In addition, EPRI reevaluated EPA's analyses of the plants found to pose the highest risks.

Power-Generation Scenarios

The 1990 Clean Air Act Amendments required that EPA project risks from U.S. power plants as they will be configured following compliance with other sections of the 1990 Amendments. For this reason, EPA first had to estimate the configuration of the industry in 2010. Increase in demand was taken into consideration, as were the requirements for compliance with other parts of the Clean Air Act Amendments, primarily reductions in SO_2 beginning in 2000. In addition, EPA estimated the number and types of new plants to be built between 1990 and 2010 to meet the increased demand. Controls to reduce NO_x emissions were also taken into account.

Some plants operating in 1990 will be retired by 2010, so the industry configuration includes both new and currently operating plants, as well as retrofits of emission controls on many existing plants to meet the required reductions in SO_2. In addition, some fuel switching is anticipated, notably a continuing decline in the market share of oil used for power generation and an increase in market share for natural gas and coal.

The EPA and EPRI estimates of the power-generation mix in 2010 are both based on a decrease in the use of oil; data collected since these projections were made have shown that the expected decline in oil was correct. But neither organization anticipated the rapid increase in the use of natural gas. It is too soon to tell how accurate the estimates are of the mix of power plant controls in 2010, because required reductions in SO_2 and NO_x have not taken full effect.

Identification of Substances

The 1990 Clean Air Act Amendment's list of air toxics includes 189 specific substances or classes of substances. Some of these (e.g., pesticides) are not likely to be emitted from a fossil-fuel-fired power plant. Both organizations focused primarily on toxic metals known to exist at trace levels in coal and oil ash. After an initial screening assessment, EPA developed a list of 13 toxic metals to be assessed for chronic exposures and three substances of concern for acute exposure; the latter are hydrochloric acid (HCl), hydrofluoric acid (HF), and acrolein. EPRI included several other substances (e.g., antimony and selenium), but these did not affect the findings. In addition to trace metals, EPA included many organic chemicals (the full list is available in Appendix A of the EPA study) (EPA, 1994). The EPRI study contained fewer organic chemicals; these included benzene, dioxins and furans, formaldehyde, polycyclic aromatic hydrocarbons (PAHs), and toluene. A partial list of the substances evaluated by EPA and EPRI is provided in Table 1. All of the substances shown on the list were evaluated for coal-fired power plants, but if the source term data indicated that they were not likely to produce significant exposures, they were not evaluated for oil-fired and gas-fired plants.

Estimated Emission Factors

An emission rate for each applicable substance was estimated for each U.S. power plant. These source-term estimates were based on measurements taken at a

TABLE 1 Partial List of Substances Included in the EPA and EPRI Analyses

EPA Analysis	EPRI Analysis
Arsenic	Arsenic
Beryllium	Beryllium
Cadmium	Cadmium
Chromium	Chromium
Dioxins/furans	Dioxins/furans
Formaldehyde	Formaldehyde
Acid gases (HCl, HF)	Acid gases (HCl, HF)
Lead	Lead
Manganese	Manganese
Mercury	Mercury
Nickel	Nickel
n-nitrosodimethylamine	
	PAHs
Radionuclides	Radionuclides

representative sample of power plants with different types of emission controls and fuels. Because many of the samples were provided to EPA by EPRI, both organizations estimated emission rates based on essentially the same raw data.

EPA and EPRI used similar but not identical methods to assess the source term of metals from coal-fired plants. Both attempted to construct a mass balance for each pollutant based on measured concentrations in the coal, in the bottom ash removed by each control device, and in the particulates emitted. EPA used this method to develop an emission factor for each configuration that was proportional to the amount of material in the coal that was burned. EPRI used a slightly different method. EPRI observed differences in the removal efficiencies of various control devices depending on the power plant capacity. Typically, the removal efficiencies were higher for coal streams with high concentrations of the pollutant of interest. The degree to which this effect occurred depended on the specific pollutant. Figure 1 indicates the observed release rates, in pounds of particulate emissions per pound of fuel input; EPRI used these and similar data to estimate source terms.

Determining the content of organic pollutants in fuels was not as useful as for trace metals because organic hazardous air pollutants (HAPs) may be created in the combustion or control processes. For this reason, both EPA and EPRI estimated emission rates for organics based on measured emissions from plants of different configurations. For some substances (e.g., PAHs and dioxins/furans), estimating source terms was complicated by a relatively high number of measurements with concentrations below the detection limits.

Naturally occurring radionuclides, principally uranium and thorium and their

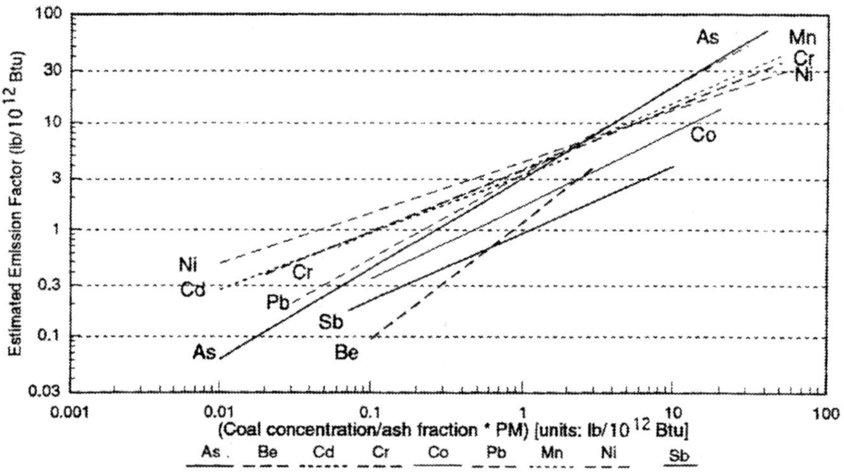

FIGURE 1 Emission factors used in the EPRI study. Source: EPRI, 1994.

decay products, are present in coal ash and in oil. EPA estimated radionuclide releases based on emission factors per unit energy consumed in generation; these estimates were developed in 1989 as part of an assessment of radionuclide emissions from coal-fired plants. In that analysis, EPA determined that controls targeting radionuclides were not necessary. The emission factors in the present study were based on the radionuclide content of the ash in the fuel (coal and oil), with adjustments made to account for nonuniform partitioning of radionuclides between fly ash and bottom ash. EPRI used a different method. EPRI based the radionuclide emission estimate on the annual mass of particulate emissions and representative concentrations of radionuclides in fly ash for the type of plant being analyzed.

Air-Dispersion Models and Exposure and Risk Assessment

The usual approach to assessing exposures to hazardous substances is to calculate an "exposure point concentration" and then apply assumptions or data regarding individual factors that would affect exposure. For the inhalation pathway, the exposure point concentration of interest is the concentration of the substances of concern in air once a plume has reached ground level. For U.S. power plants with tall stacks, a plume may reach the ground about 10 kilometers from the plant. In both the EPA and EPRI analyses, air-dispersion calculations were made using versions of EPA's integrated-source, complex, long-term (ISCLT) model.

Given a calculated exposure point concentration over time, an individual exposure assessment includes many factors, such as an individual's inhalation rate (which varies with the level of activity), the mix of time spent indoors and outdoors (indoor concentrations tend to be lower), the length of time an individual remains at the location of interest, and body weight (when risk factors are based on intake rates expressed as microgram (μg)/day per kilogram of body weight). In addition to individual exposure rates, these studies also estimated health effects for the total exposed population.

The substances of concern in these studies included both carcinogens and noncarcinogens. For carcinogens, it is standard practice in the United States to assume a linear relation between exposure and risk, with no threshold for low doses or low dose rates. This requires using a risk factor in which risk is proportional to intake relative to body weight. A consequence of this assumption is that the risk to an individual is directly proportional to the individual's cumulative lifetime intake of a carcinogenic substance. Similarly, the cancer risk in a population, in terms of the expected number of additional cases per year, is assumed to be proportional to the total exposure of the population.

For noncarcinogens, EPA limits exposures based on a defined level called a reference dose (RfD); for inhalation exposures, a reference concentration (RfC) is sometimes used. EPA considers exposures at these levels to be virtually risk free, but exposures above the RfD or RfC level is to be avoided. Exposures to noncarcinogens are expressed in terms of a dimensionless index called a hazard quotient (HQ), which is simply the ratio of exposure to the RfD. For example, the RfD for methylmercury is 0.1 μg/kg-day (one microgram per day per kilogram of body weight). For a person who weights 70 kg, the reference dose is 7 μg/day. This means that a person who weighs 70 kg who consumes 0.7 μg/day of methylmercury would have an HQ of 0.1. If there are exposures to more than one substance, the HQs are added up to determine a hazard index (HI).

EPA and EPRI used the same toxicity values for the substances studied. The cancer unit risk factors, RfDs, and RfCs were based on the current values (at the time the studies were conducted) on EPA's Integrated Risk Information System (IRIS) (EPA, 1994).

Results

The risk assessments for inhalation were generally very low. Figure 2, which shows the results of the EPA study for coal-fired plants, indicates that the calculated lifetime cancer risk to the maximally exposed individual exceeded one in a million for only two plants. As a point of comparison, EPA frequently uses an acceptable risk range for lifetime cancer risks of one in ten thousand to one in a million. Risks of less than one in a million are judged to be too low to be of concern. As Figure 3 shows, the inhalation cancer risk comes primarily from arsenic and chromium, with only a small contribution from cadmium. Radionuclides are not included in this analysis.

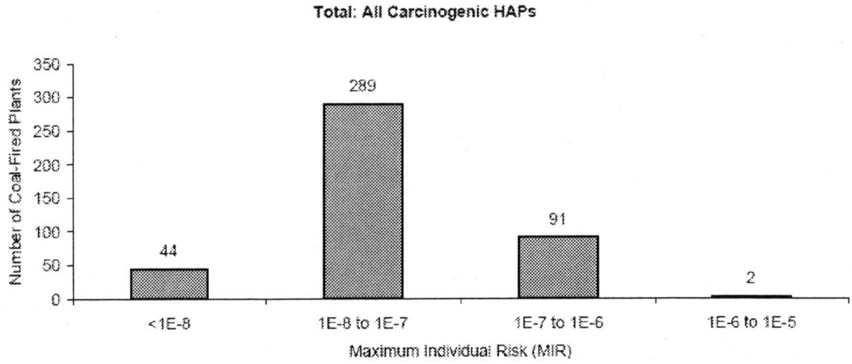

FIGURE 2 Maximum individual risk (MIR) of cancer from coal-fired plants. Source: EPA, 1998.

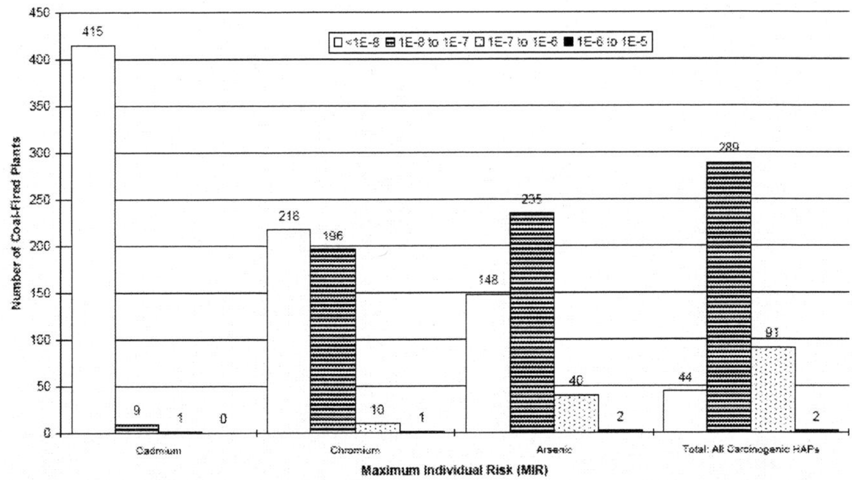

FIGURE 3 EPA results of inhalation cancer risks by chemical for coal-fired power plants. Source: EPA, 1998.

Surprisingly, both EPA and EPRI found that many of the plants that pose the highest risk burn oil. This may be because many oil-fired plants are small to medium-sized peaking plants in urban locations with generally shorter stacks than coal-fired plants and with no controls for particulate emissions. Figure 4 shows EPA's results for oil-fired plants. The primary contributor to inhalation cancer risk from the oil-fired plants comes from nickel. We must keep in mind that this finding depends on the toxicity assumptions applied to nickel. EPA has

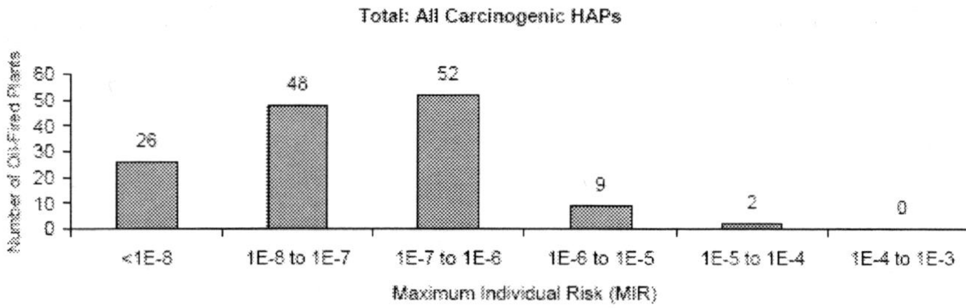

FIGURE 4 EPA results for oil-fired power plants. Source: EPA, 1998.

not classified soluble nickel salts as carcinogens but has classified nickel refinery dust, nickel carbonyl, and nickel subsulfide as carcinogens. Both EPA and EPRI found that the risks from gas-fired plants were low. As Figures 5 and 6 show, EPRI's analytical findings for inhalation cancer risks were consistent with EPA's findings. Figure 7 shows that the range of calculated risk for each plant type varied widely, depending on the location, stack characteristics, weather conditions, local population, and whether the setting was urban or rural.

As for noncancer end points, both studies found that exposures were well below the applicable RfDs and RfCs. Figure 8 shows cumulative distributions of the HIs calculated by EPRI. The HI for coal plants is due mainly to HCl and manganese.

Both EPA and EPRI evaluated multipathway exposures, especially for substances where the main concern was not inhalation exposure but accumulation in the food chain (mercury, dioxins and furans, and others). In addition, the studies took into account indirect exposures to arsenic, lead, cadmium, and radionuclides. A multipathway risk assessment is much more complex and requires much more input data than an inhalation risk assessment. The EPA study noted that power-plant emissions of lead and cadmium are roughly 1 percent of the total annual U.S. emissions of these metals and, therefore, concluded that the incremental exposures from power-plant releases were small in comparison to releases from other sources.

Mercury

Mercury has long been identified as a neurotoxin. As poisoning episodes at Minimata, Japan, and in Iraq have demonstrated, high-dose exposures to mercury can cause severe damage to human health. In the context of coal-fired power plants, the concern is that mercury releases to the environment will be deposited on watersheds, migrate into lakes and rivers, be converted to methylmercury in lake and river sediments, and accumulate in fish.

In 1994, when the EPRI study was published, the RfD for exposure to

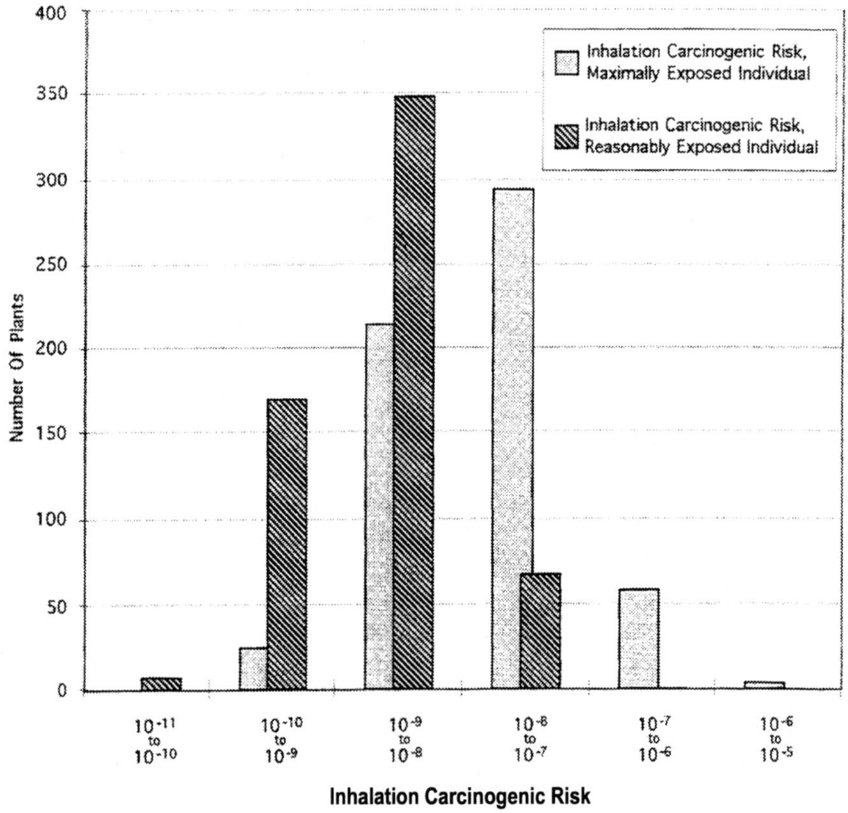

FIGURE 5 Results from the EPRI inhalation risk assessment. Source: EPRI, 1994.

methylmercury was based on a dose high enough to cause parasthesia (numbness and a tingling sensation in the extremities, typically the fingers). This RfD was 0.3 μg/kg-day (that is, the defined safe intake rate was 0.3 μg/day per kg of body weight). In 1997, when the EPA study was published, a revised RfD of 0.1 μg/kg-day was used, based on developmental effects observed in connection with methylmercury poisoning in Iraq. Neither RfD was based on exposure through fish consumption.

Since the publication of the EPA electric utility air toxics study, EPA has updated the methylmercury RfD based on epidemiological studies of populations that consume large amounts of fish (relative to the United States). Evaluations of children's performance on a variety of childhood development tests showed that poor performance was correlated with high maternal exposure to methylmercury, indicated by measurements of methylmercury in the mother's

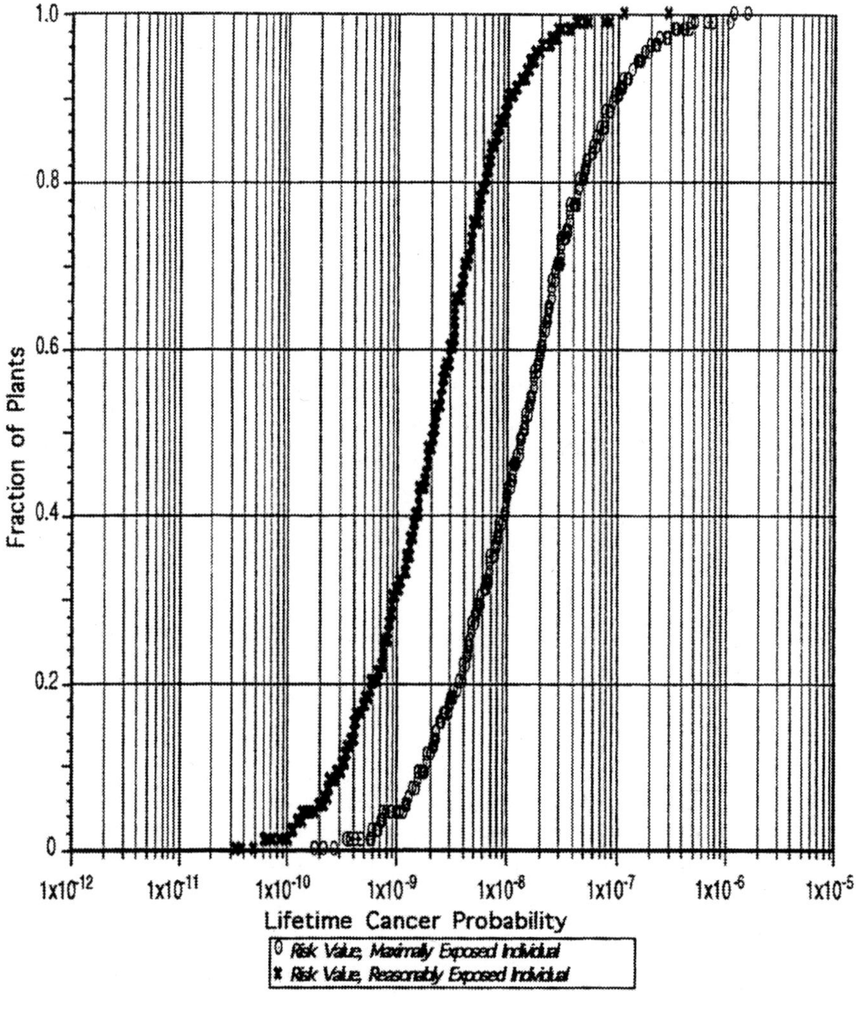

FIGURE 6 Cumulative distribution of inhalation risks from the EPRI analysis. Source: EPRI, 1994.

hair or blood or by a dietary survey concerning fish consumption. Although the current basis for the RfD is concerns about exposures during pregnancy, the RfD is the same, 0.1 µg/kg-day.

In EPA's case studies, exposures to methylmercury from power plants alone did not exceed the RfD for the typical recreational angler (assumed to consume 8 grams per day of freshwater fish). However, the RfD would be exceeded for someone higher on the consumption distribution, such as a subsistence fish

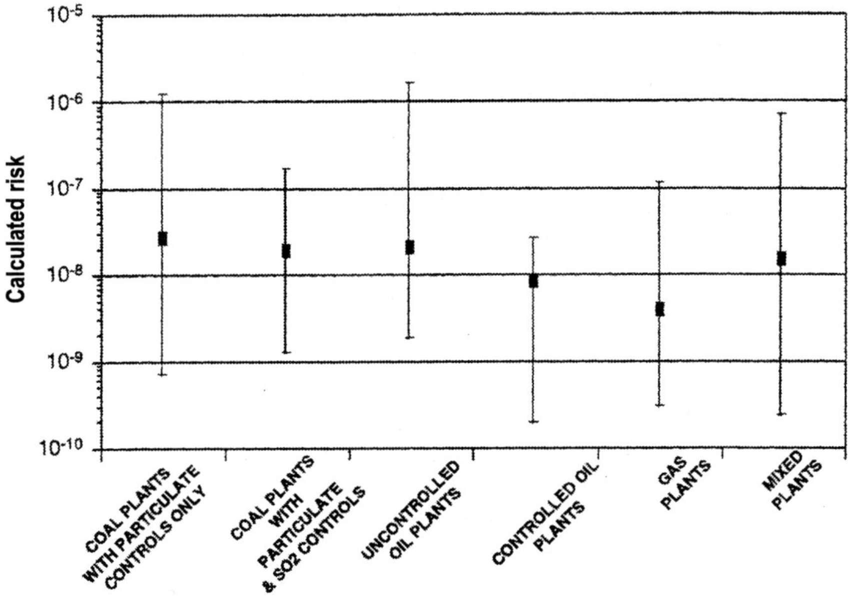

FIGURE 7 Comparison of inhalation cancer risk by plant type. Source: EPRI, 1994.

consumer. Although mercury exposures occur mainly through the consumption of marine fish, incremental exposures from power-plant emissions added to other exposures could exceed the RfD for a significant number of people.

In the EPRI case studies, the highest calculated methylmercury exposure led to a calculated HI of 0.28, based on an RfD of 0.3 μg/kg-day. This corresponds to an HI of 0.84 based on the current RfD. The HI was calculated for the contribution from each case-study plant and did not include exposures from other sources, such as the consumption of marine fish.

Arsenic

EPA and EPRI both evaluated potential exposures to arsenic through pathways other than inhalation. The focus was on inorganic arsenic, which is believed to be much more toxic than organic arsenic. This distinction is important because arsenic can accumulate in fish and shellfish, but usually in the organic form. Inorganic arsenic deposited on soil that can be taken up in crops is of greater concern. With the exception of high exposures and calculated risks to children

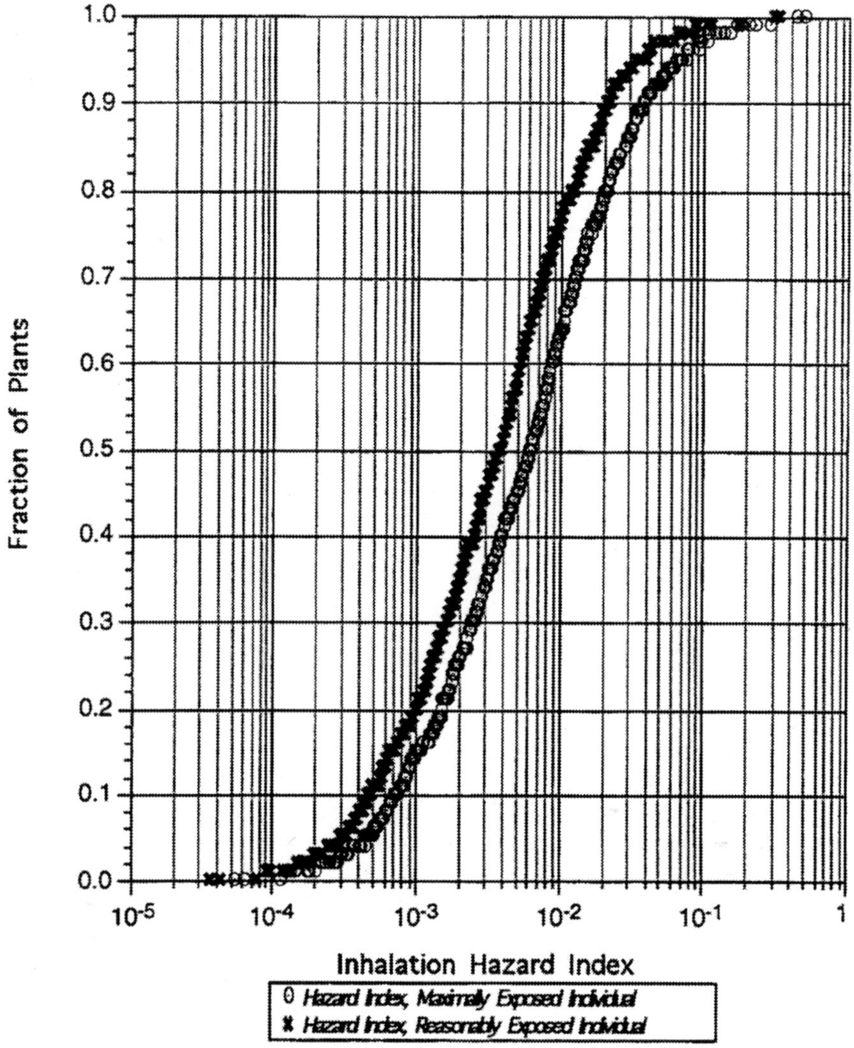

FIGURE 8 Inhalation hazard index for all plants. Source: EPRI, 1994.

with pica (a disorder that causes them to eat large amounts of dirt), the lifetime cancer risks from arsenic were typically in the range of one in a million or less. EPA further noted that, even after 30 years of accumulated deposition, the arsenic contributions to soils near power plants would be around 10 percent of average background concentrations. Thus EPA concluded that power-plant emissions make a relatively small contribution to arsenic exposure in the United States.

Radionuclides

Prior to the study mandated by the Clean Air Act Amendments of 1990, EPA had evaluated the risks from, and need for regulation of, radionuclide emissions from coal-fired power plants at roughly five-year intervals going back to 1979. Each time, EPA had concluded that regulation was not necessary to protect public health. However, EPA had set NESHAPs for emissions of radionuclides from other sources, such as phosphate mines and U.S. Department of Energy sites, with a limitation on exposure of 10 millirems (mrem) per year (0.1 millisievert [mSv] per year).

In the assessments of radionuclide risks from power plants, EPA used CAP93-PC, a modified version of CAP88-PC. Both versions were designed to assess doses and risks to individuals and populations from specified releases of radionuclides to the atmosphere. The model includes a version of the Industrial Source Complex model used by EPA and EPRI for the inhalation analyses. With local meteorological data, CAP93-PC is capable of modeling emissions from power plant stacks. The model can perform calculations of exposures through inhalation, through ingestion of food, through incidental consumption of soil, through drinking water, through external radiation, through immersion in radionuclides in air, and through radionuclides deposited on the ground.

EPA ran this model for all U.S. plants, considering emissions of uranium and uranium decay-chain radionuclides, thorium and thorium decay-chain radionuclides, and potassium-40. These analyses resulted in the highest risk estimates in EPA's study; 17 plants were estimated to produce exposures resulting in maximum individual lifetime risks greater than 1×10^{-5}. For each of these 17 plants, the dominant exposure pathway was external radiation from radionuclides deposited on the ground surface through wet deposition from the plume. Because the concentrations in the plume are highest near the stack, ground concentrations were also calculated to be highest near the stack. Given the number of plants that had to be assessed, it was not possible to get actual data for the actual locations of the nearest residents around each plant. In general, the risk reported was for a person living 200 meters from the power plant stack in the direction of the prevailing wind.

The U.S. power plant industry, through the Utility Air Regulatory Group (UARG), conducted a reanalysis of the top 17 plants in EPA's study using the actual location of the nearest neighbors, typically determined by aerial photographs of the plant and surrounding area. Most of the plants were in rural locations, many with no residents living within a few thousand meters of the plant. UARG's review revealed that in several cases the point of maximum risk in EPA's analysis was inside the plant boundary, for example, on a coal pile or cooling pond. In discussions between the industry and EPA, it was agreed that the EPA analyses were conservative, but because EPA showed no interest in regulating radionuclide emissions, it was not considered important to get more refined and accurate risk estimates.

EVENTS SINCE THE STUDIES WERE PUBLISHED

Based on the air toxics study and the companion mercury study report to Congress, EPA concluded that, with the exception of mercury emissions, power plant emissions of toxic substances did not require regulation to protect public health or the environment. However, because power plants are the largest sources of anthropogenic mercury emissions in the United States, EPA concluded that controls to reduce mercury emissions from power plants are justified. EPA is in the process of determining the limitations on mercury in power-plant emissions. In EPA's analyses of mercury exposures, data on fish consumption, when combined with data on the mercury concentration in fish, indicated that a significant number of people in the United States were exposed to methylmercury at higher levels than the EPA RfD. Most of this exposure was from the consumption of commercially purchased marine fish, but some people were thought to be overexposed through the consumption of self-caught freshwater fish. The mercury content of marine fish is primarily from global background concentrations of mercury and would not be affected by changes in U.S. power plant emissions. There is a stronger relationship between U.S. emissions and the concentration of methylmercury in freshwater fish.

At the time these studies were conducted, exposure limits for methylmercury were based on an episode of acute poisoning in Iraq, in which methylmercury used as a fungicide on seed grain was accidentally consumed. Because these were high exposures of short duration, some people in the toxicology community were concerned that they were not a sufficient basis for setting exposure limits for the consumption of methylmercury in fish. In addition, all of the studies of mercury exposures in populations that consume large amounts of fish have been found to have significant limitations. Two new large epidemiological studies, one in the Seychelles and one in the Faeroe Islands, to evaluate the neurological development of children born to mothers who consume large amounts of methylmercury were under way to provide better information. In response to a congressional requirement, EPA funded a study by the National Academies of health risks associated with methylmercury exposure (NRC, 2000). The study supported the continued use of the methylmercury RfD of 0.1 μg/kg-day, based largely on the results of the Faeroe Islands study. The population of most concern is pregnant women.

Another data gap at the time of the studies was actual exposure levels to methylmercury. Exposures were based on food diary studies coupled with measured concentrations of methylmercury in fish tissue. The primary diary study was the Continuing Study of Food Intake by Individuals (SCFII), a three-day diary study conducted by the U.S. Department of Agriculture. Unfortunately, a three-day study is not a good way to estimate the upper tail of the distribution curve for consumption of fish, which is eaten infrequently. This is particularly true for wild freshwater fish, which Americans consume, on average, less often than marine fish and which is often not included in food statistics because it is not bought or sold commercially. As a result, uncertainties regarding the upper tail of the exposure distribution were high, with the largest uncertainties associated with exposures from the consumption of freshwater fish. Farm-raised fish are not sources of significant methylmercury exposure because they are not subject to the normal bioaccumulation processes.

An alternative to fish consumption surveys is biomarker data because methylmercury exposures can be determined through measurements of mercury concentrations in hair or blood. Data released in June 2002 from the federal National Health and Nutrition Examination Survey (NHANES) provided an opportunity to evaluate blood levels of mercury in 1,709 women of childbearing age and 705 children between the ages of one and five. As part of the study, the women were also asked about their fish consumption.

The NHANES study results, when weighted to adjust for demographic factors, indicated that about 7.7 percent of women in the United States aged 16 to 49 have blood levels of methylmercury above the RfD. The fish consumption survey of the NHANES study did not have the resolution necessary to determine the contribution of wild freshwater fish to exposures at the upper tail of the exposure distribution.

REFERENCES

EPA (Environmental Protection Agency). 1994. Integrated Risk Information System. Available online at: *http://www.epa.gov/iris/*.

EPA. 1997. Mercury Study Report to Congress. R-452/R-97-003. Washington, D.C.: Office of Air Quality Planning and Standards and Office of Research and Development, Environmental Protection Agency.

EPA. 1998. Study of Hazardous Air Pollutant Emissions from Electric Utility Steam Generating Units: Final Report to Congress. EPA-453/R-98-004a. Washington, D.C.: Office of Air Quality Planning and Standards, Environmental Protection Agency. Also available online at: *http://www.epa.gov/ttn/oarpg/t3/reports/eurtc1.pdf*; the appendices are available at *http://www.epa.gov/ttn/oarpg/t3/reports/eurtc2.pdf*.

EPRI (Electric Power Research Institute). 1994. Electric Utility Trace Substances Synthesis Report. Volumes 1–4. EPRI TR-104614-V1 through V4. Palo Alto, Calif.: Electric Power Research Institute.

NHANES (National Health and Nutrition Examination Survey). 2000. NHANES 1999–2000: Data Files, Documentation/Codebook and SAS Code. Available online at: *http://www.cdc.gov/nchs/about/major/nhanes/NHANES99_00.htm*.

NRC (National Research Council). 2000. Toxicological Effects of Methylmercury. Washington, D.C.: National Academies Press.

Philadelphia Department of Public Health. 2004. History of Air Pollution Control in Philadelphia. Available online at: *http://www.phila.gov/health/units/ams/pdf/History.pdf* (accessed May 2004).

Environmental Performance of Coal-Fired Power Plants Financed by the World Bank

JACK J. FRITZ
National Academy of Engineering

This paper provides a summary of the findings from an environmental performance review of coal-fired power plants in China constructed between 1986 and 1997 and financed through World Bank lending operations. This review, carried out in 2000, included visits to six plants (600 to 1,200 megawatts [MW] each, Figure 1), an analysis of plant survey data, and recommendations for improving the environmental performance of coal-fired power plants in general (Jia et al., 2000). To date, this is the only analysis of its kind carried out in China.

BACKGROUND

Since the mid-1980s, the World Bank has assisted China in developing its electric-power industry, first by financing thermal power plants and hydropower plants, transmission and distribution systems, and renewable energy systems, and second by supporting policy initiatives to make the electrical energy sector more autonomous, environmentally sustainable, and financially viable. Since 1986, the World Bank has financed approximately 9,000 MW of coal-fired generation at an investment cost of $9.4 billion; the World Bank's contribution was $2.9 billion. The plants were built to modern standards of operation and pollution control.

From 1986 to 1997, environmental standards, guidelines, and enforcement institutions changed significantly. As increasingly prosperous urban residents clamored for improvements in air quality, environmental management became a major public policy issue. The power sector is often accused of being the biggest source of urban air pollution, but most ground-level particulate and sulfur

FIGURE 1 Coal-fired power plants visited by mission (☆) and coal-fired power plants under construction, not visited by mission (○).

dioxide (SO_2) emissions originate not from the power sector but from thousands of coal-fired boilers that are ubiquitous in most urban areas in China.

Coal continues to be the predominant energy source in China, which is the world's largest producer and user of coal. China uses approximately 1.4 billion tons of coal annually, and this figure continues to rise as primary energy consumption rises (5.4 percent annually from 1980 to 1996). In 1996, for example, coal accounted for 74.8 percent of total commercial energy production in China. The industrial sector is the largest consumer of coal at 68.5 percent of final energy consumption. With recoverable coal reserves estimated at 1,009 billion metric tons, China will certainly depend on coal as its main energy source for some time to come.

China is the second largest power producer in the world after the United States, and from 1980 to 1997, both countries increased their generating capacity significantly. Electricity generation in China grew at an annual rate of 8.9 percent, reaching 254 gigawatts (GW) and producing 1,134 terawatt hours (TWh) annually. Since 1988, roughly 11 to 15 GW of generating capacity has been added annually. As a result, by 1997 most power shortages had been eliminated.

Of the total installed generating capacity of 254 GW as of 1997, roughly 76 percent came from the burning of fossil fuels, 22 percent came from hydropower, and roughly 2 percent came from nuclear power. Although the government has put a high priority on hydropower development, so far its efforts have been constrained by the great distances between the resources and the primary load centers (generally more than 1,500 kilometers). The rationale behind the controversial Three Gorges Project, now nearing completion, is based on the large scale of this energy source and expected flood-control benefits. In 1997, coal-fired power plants provided 82 percent of total electricity generation. The proportion of electricity provided by oil-fired plants has been declining since the 1980s because of falling production in the country's aging oil fields. Gas supplies are only now coming on line, primarily for urban household use in Beijing. Nuclear power, which currently accounts for about 2.1 GW, is emerging as a significant source of electrical power.

China, with 28 provincial and three large municipal power grids, has been actively pursuing system interconnection to improve efficiency and alleviate power shortages in the major load centers. When the government was restructured in 1998, the Ministry of Electricity was replaced by the State Power Corporation (SPC), which has emerged as the responsible agency for the power sector throughout China.

WORLD BANK LENDING OPERATIONS

The World Bank has financed nine coal-fired power plants in China since 1985. The first project was Beilungang #1, near Hangzhou. The most recent was Wai Gao Qiao in Shanghai, with a capacity that will eventually reach 5,000 MW. Total capacity is about 10,000 MW, with a total investment of $9,413,000 ($2,920,000 from the World Bank) (Table 1). The purpose of assisting China has been to satisfy the growing demand for electric power, improve efficiency, and reduce adverse environmental impacts. The leaders of China are well aware of the link between economic development and the ready availability of electrical energy. In fact, China has pursued an active policy of rural electrification for decades.

China's overwhelming dependence on coal for power generation has serious environmental consequences at every stage—extraction, transport, and power generation. The government has established a number of targets for stabilizing pollution at 1995 levels: (1) to restrict the discharge of particulates to 3.8 million tons annually for all coal-fired plants with more than 6 MW capacity on any one grid through the use of electrostatic precipitators with average collection efficiencies of 98 percent; (2) to restrict the discharge of SO_2 by coal-fired power plants on any one grid to 6.5 million tons annually; (3) to adopt low-nitrogen oxide (NO_x) technology for all new plants of 300 MW or more; (4) to ban the discharge of coal ash into waterways and recycle at least 40 percent (45 million tons) of fly ash annually; and (5) to require that 70 percent of wastewater from power plants meet national discharge standards.

TABLE 1 World Bank Lending for Thermal Power Plants in China (in thousands of U.S. dollars)

No.	Name	Location	Date	Capacity (MW)	Total	World Bank Investment
1	Beilungang I	Zhejiang	May 1986	600	$1,044.09	$225.00
2	Beilungang II	Zhejiang	May 1987	600	$289.70	$165.00
3	Wujing	Shanghai	February 1988	2 × 300	$354.10	$190.00
4	Yanshi	Henan	December 1991	2 × 300	$459.60	$180.00
5	Zouxian	Shandong	March 1992	2 × 600	$957.40	$310.00
6	Beilungang II	Zhejiang	March 1993	2 × 600	$1,350.00	$400.00
7	Yangzhou	Jiangsu	February 1994	2 × 600	$1,081.40	$350.00
8	Tuoketuo	Inner Mongolia	April 1997	2 × 600	$1,300.50	$400.00
9	Wai Gao Qiao	Shanghai	May 1997	2 × (900 to 1,000)	$1,898.00	$400.00
10	Leiyang	Hunan	March 1998	2 × 600	$678.00	$300.00
	TOTAL			10,200 to 10,400	$9,413.60	$2,920.00

Environmental goals are also being pursued through electricity pricing reforms aimed at increasing cost consciousness and strengthening the enforcement capabilities of provincial environmental protection bureaus. The government also encourages utilities to locate power plants near mine mouth operations, to develop plants with a high heat rate, to retrofit older, smaller plants with new combustion technologies, to adopt coal washing and beneficiation technologies, and to adopt flue-gas desulfurization for plants that use medium- and high-sulfur coals. The government has also launched a $2 billion program to control sulfur emissions from a wide variety of sources, including the many small boilers scattered throughout Chinese cities.

EVOLUTION OF CHINA'S STANDARDS AND GUIDELINES

The first emission standards for coal-fired thermal power plants in China, applied in 1992 (GB 13223-91), were limited to emissions of particulates and SO_2 from existing and modified plants. In 1997, the standards were updated and expanded to include emissions of NO_x (GB 13223-96). Under these updated standards, power plants fall into one of three categories, depending on when approval was granted for original construction: Phase I, approved before August 1, 1992; Phase II, approved between August 1, 1992, and December 31, 1996; and Phase III, approved after January 1, 1997.

As Table 2 shows, the old and new requirements for Phase I and II plants did not change. Major changes for Phase III plants include: (1) regulation of dust emissions independent of coal ash content; (2) criteria for both mass flow and concentration of SO_2 in flue gas; and (3) standards for NO_x emissions. Although the 1992 standards effectively addressed particulate emissions (some 90 percent of existing or upgraded plants, as well as new plants, have installed or retrofitted electrostatic precipitators), air quality in many Chinese cities has continued to deteriorate, necessitating tighter standards for SO_2 and NO_x. Recent interest has focused on the relationship of PM_{10} (fine particulate matter with a diameter of less than or equal to 10 micrometers) to total suspended particulates (TSP). Updates may have been made since 1997.

Standards for liquid effluents were established for all industries in 1988 (GB 8978-88) and modified in 1996 (GB 8978-96). Clearly, liquid effluents are less of a problem for thermal power plants than for industries and municipalities. Major effluents from power plants include runoff from coal piles, sanitary wastes, chemical wastes from various subsystems (e.g., boiler blow down), and cooling water. Typically, much of the cooling water and chemical wastewaters are recirculated, partly because the dilution of wastewater to meet the standard is prohibited. The effluent standards for power plants are the same as for industrial and municipal discharges, a five-step system that classifies ambient water quality according to the ultimate use of the receiving water (e.g., drinking, industrial, agricultural, etc.).

TABLE 2 Chinese National Standards for Flue-Gas Emissions from Coal-Fired Power Plants

Pollutant	GB 13223-91			
	Plant Type			
	Existing Power Plant		New, Extended, Reconstructed	
	ESP	Other Dust Control	Power Plant > 670 t/h or Urban[b]	< 670 t/h and Urban[b]
Particulate (mg/Nm^3)	200–1,000[a]	800–3,300[a]	150–600	500–2,000
Sulfur dioxide-mass flow (t/h)	Correlation based on average wind speed and stack height			
	—			
Sulfur dioxide-concentration (mg/Nm^3)	—	—	—	—
Nitrogen oxides (mg/Nm^3)	—	—	—	—

[a]Range reflects ash content of coal. Lower value is < 10 percent; higher value is > 40 percent; intermediate values are specified in the regulation.

[b]Value depends on boiler size (coal feed rate in t/h) and/or regional characteristics.

[c]Range reflects regional characteristics: lower value is for urban area; upper value is for rural area.

EVOLUTION OF WORLD BANK GUIDELINES

The evolution of World Bank guidelines goes back to the 1970s, but they were not compiled and made a formal requirement until 1984. Initially, these guidelines were not the ironclad requirements they have become of late. The initial emphasis was on controlling emissions of particulates, SO_2, and NO_x from any combustion process. The guidelines include two types of standards: mass-based standards (e.g., tons per day) and concentration-based standards (e.g., milligrams per normal cubic meter [mg/Nm^3]). Mass-based emissions can be used to calculate concentrations of SO_2 in flue gas using the heat rate (gram coal equivalent per kilowatt hour [gce/kWh]) and flue-gas flow (e.g., 350 Nm^3/gigajoule [GJ], with 6 percent excess O_2). As Table 3 indicates, the 1984 guidelines for particulates were based on concentration; for NO_x they were based on mass per unit heat release; and for SO_2 they were based on mass.

In 1997, guidelines were written specifically for thermal power plants. At that time, World Bank membership was expanded to include the countries of

GB 13223-96				
Phase				
I		II		III
ESP	Other Dust Control	> 670 t/h or Urban[b]	< 670 t/h and Urban[b]	Urban or Rural
200–1,000[a]	800–3,300[a]	150–600	500–2,000	200–600[c]
Correlation based on average wind speed and stack height				
—	—	—	—	1,200–2,100[d]
—	—	—	650–1,000[e]	

[d]Range reflects sulfur content of fuel: lower value is for coal containing > 1.0 percent; higher value is for coal containing < 1.0 percent.
[e]Range reflects boiler type: lower value is dry bottom; upper value is wet bottom.
Source: Jia et al., 2000.

Eastern Europe and the former Soviet Union, where the power sector required tighter, but more flexible, guidelines. The basic differences between the 1984 and 1997 guidelines are: (1) tighter control of particulates (down to 50 mg/Nm^3) and (2) a mass-based standard for SO_2, but with a ceiling for both mass and concentration. The new standards provide more flexibility in the choice of coal but reduce emissions overall.

AMBIENT AIR QUALITY TRENDS AND STANDARDS

Monitoring data from 1991 to 1998 for 60 medium-sized and large cities in China showed some improvements in ambient air quality. Forty cities experienced reductions in TSP, and 50 had reductions in SO_2. The median concentration of SO_2 in 32 cities with populations of more than one million dropped from 100 to 62 μg/m^3. In smaller cities, the average concentration dropped from 50 to 32 μg/m^3. Anecdotal evidence suggests, but does not prove, equivalent declines in health-related damage based on typical linear dose-response methodologies.

TABLE 3 World Bank Guidelines for Air Pollution from Thermal Power Stations

Pollutant	1984 Guidelines	1997 Guidelines[b]
Dust		
Emissions (mg/Nm^3)	100–150[a]	50
Ambient ($\mu g/Nm^3$)		
Daily maximum	500	—
Annual average		
Sulfur Dioxide		
Emissions (tons/day)	100–500[a]	$0.2/MW_e$ for first 500 MW_e $0.1/MW_e$ for subsequent capacity (maximum 500 tons/day) and < 2000 mg/Nm^3
Ambient ($\mu g/Nm^3$)		
Daily maximum	500	—
Annual average		
Nitrogen Oxides		
Emissions (nanograms/joule)	260 (lignite) 300 (coal)	260 (coal) 1,500 (low volatile coal)
Ambient ($\mu g/Nm^3$)		
Annual average	100	—

[a]Allowable value for ambient levels; the lower the ambient level, the higher the allowable emission level.

[b]Updated World Bank guidelines do not provide ambient values.

TSP levels, however, remained fairly high, with only a few decreases in some urban areas. The median concentration in China's 32 largest cities dropped from 334 to 332 $\mu g/m^3$; in smaller cities, it declined from 260 to 215 $\mu g/m^3$. In some large cities, such as Tianjin and Beijing, which had significant increases in population, TSP increased significantly. With the recent introduction of natural gas to Beijing, levels of both TSP and SO_2 are expected to decrease appreciably.

Ambient air quality standards were promulgated in 1982 (GB 3095-82) and updated in 1996 (GB 3095-96) (see Table 4). The standards include PM_{10} based on epidemiological evidence suggesting that much of the health damage from air pollution is caused by exposure to fine particles. Ambient air quality standards are based on an air-quality classification system: Class I areas are natural conservation districts, resorts, and tourist areas of historic interest; Class II areas include urban, residential, commercial, and rural areas; Class III areas are industrial areas or areas with high traffic volumes.

A key aspect of the 1996 makeover was an increase in the time and coverage of monitoring. Under the original regulatory regime (prior to 1996), the following protocol was standard:

TABLE 4 Chinese Ambient Air Quality Standards (μg/m^3)

	GB 3095-82			GB 3095-96		
	Air Shed Classification			Air Shed Classification		
Pollutant	I	II	III	I	II	III
TSP[a]						
Annual average	—	—	—	80	200	300
Daily maximum	150	300	500	120	300	500
One-time maximum	300	1,000	1,500	—	—	—
PM[b] <10 μ						
Annual average	—	—	—	40	100	150
Daily maximum	50	150	250	50	150	250
One-time maximum	150	500	700	—	—	—
Sulfur Dioxide						
Annual average	20	60	100	20	60	100
Daily maximum	50	150	250	50	150	250
One-time maximum	150	500	700			
Nitrogen Oxides						
Annual average	—	—	—	50	50	100
Daily maximum	50	100	150	100	100	150
One-time maximum	100	150	300	—	—	—

[a] Total suspended particulates.
[b] Particulate matter.

- four monitoring campaigns per year (one every three months)
- five- to seven-day monitoring campaigns
- four six-hour intervals per day
- 15 to 20 minutes of monitoring per interval

At best, the original protocol provided monitoring data for 37 hours per year, or a sample window of 0.43 percent, hardly a statistically relevant sample.

Because the World Bank required background ambient air quality data for at least a year prior to construction, an entirely new approach was necessary. Under the monitoring requirements instituted in 1996, the sampling frequency was increased:

SO_2/NO_x
- 144 days per year
- 18 hours per day
- 45 minutes per hour

TSP/PM_{10}
- 60 days per year
- 12 hours per day
- 45 minutes per hour

Thus, the sampling window increased by a factor of 52 for SO_2 and NO_x and a factor of 14.5 for TSP and PM_{10}. The new sampling procedures provided more representative data and reduced the barrier to further investment in the power sector by the World Bank and other donors.

Standards for noise levels were also enacted, as well as standards for the thermal pollution of receiving waters. Finally, in 1996, the Ministry of Electric Power (the forerunner of SPC) issued its own regulations for environmental monitoring. In addition to the new standards and guidelines, the published documents described the management, budget, equipment, reporting, compliance, and mitigation by each plant. All power plants were required to comply within six months; thus, today we have a rich database for the power sector.

THE POWER PLANT REVIEW

Objectives

The power plant review (Jia et al., 2000) was undertaken for several reasons. First, the World Bank had never carried out a comprehensive environmental review of its investment in the thermal power plant sector in China. Thus, one objective was to provide an assessment of compliance with both Chinese and World Bank environmental guidelines. In addition, the review provided an assessment of (1) adherence to an environmental management plan governing daily power plant operations and (2) the overall impact of environmental oversight of the power sector as a whole.

In addition to a review of day-to-day environmental performance, the review provides an ex post facto assessment of compliance with environmental management plans, as described in the documentation required by the World Bank and the Chinese government, which was a condition of the loans in all cases. The environmental management plans describe a host of measurements, standards, and procedures that must be followed in operating plants; these are generally considered standard operating procedures in most developed countries. The SPC was also interested in using the results to provide guidance to provincial power authorities with power plants in the pipeline on compliance with environmental guidelines in the future.

Basic Findings

The review team of three engineers visited six of the nine plants financed by the World Bank during the 1990s. These were state-of-the-art, large-scale units equipped with particulate and sulfur controls (or they used low-sulfur coals), as well as treatment systems for discharged cooling water, sanitary water, and coal pile runoff. Table 5 shows the basic characteristics of the six plants. Most of the plants the team visited were in compliance with both World Bank and Chinese guidelines as applied at the time of construction and initial operation. Plants built after 1994, primarily use high-efficiency precipitators and low-sulfur coals, and have improved overall designs. Unfortunately, older plants continue to use high-sulfur coal with less control. Industrial boilers, which also continue to use high-sulfur coal, generally have lower stacks, and are much smaller than power plants, tend to spread particulates and SO_2, mostly in heavily urbanized areas.

A clear benefit of the involvement of the World Bank has been to encourage environmental management of overall plant operations (including the treatment of cooling water and wastewater) at other plants in the same power grid. In some cases, older units have been retrofitted with precipitators that meet both Chinese and World Bank standards. The World Bank's continued policy advice and emphasis on incremental environmental improvement, and the resulting use of more efficient power technologies, have led to a much more efficient power sector than would have been possible otherwise. Table 6 outlines some of the positive impacts on China's power sector.

The World Bank has also promoted a vigorous social agenda that includes public participation, consultation, and compensation for assets appropriated by the state for plant sites. The essence of this policy is to use "resettlement action plans" to ensure that affected parties are not unduly negatively impacted. To implement this policy, the World Bank and the Chinese government developed a rather complex process that includes compensating people forced to give up land and providing information to local residents about expected environmental

TABLE 5 Basic Characteristics of Power Plants

Basic Characteristics	Beilun I	Wujing	Yanshi	Beilun II	Zouxian	Yangzhou
Unit capacity (MW)	600	300	300	600	600	600
Unit number	2	2	2	3	2	2
Stack height (m)	240	210	240	240	240	240
Ash content (%)	14.7%	7.9%	30.6%	15.5%	22.4%	10.9%
Sulfur content (%)	0.85%	0.62%	0.81%	0.63%	0.70%	0.36%
Heat value (kJ/kg)	22966	22474	17201	22404.8	22535	21610
ESP efficiency (%)	99.80%	99.80%	99.20%	99.90%	99.60%	99.80%
Heat rate (gce/kWh)	307	332	350	300	320	337
Plant efficiency (%)	40.1%	37.1%	35.2%	41.0%	38.4%	36.5%
Flue-gas volume (Nm^3/kg)	8.0381	7.8659	6.0203	7.8417	7.8873	7.5635
Coal consumption (t/d)	11265	6224	8573	16925	11966	13141

TABLE 6 Positive Impacts of World Bank Involvement

Technology Transfer	Institutional Efforts
First 600 MW subcritical unit in China	Preparation for possible privatization of some components of the power system.
First supercritical 900/1,000 MW unit in China	Development of Chinese environmental assessment procedures for power plants that parallel World Bank policies.
Low-NO_x burners	
Continuous environmental monitoring equipment	New national standards for ambient air quality that include requirements for fine particulates (PM_{10}).
High-efficiency electrostatic precipitators and flue gas desulfurization	New emission standards that include requirements for NO_x and a commitment to sulfur-control measures for coal with more than 1 percent sulfur.
Pilot test of tradable sulfur emissions	

impacts. The six plants visited for the review had followed the long-standing Chinese tradition of providing compensation, but public participation was not always solicited in an open and helpful way. In some cases, local resistance had been quieted by the perception that the project would provide jobs. Crews from all six plants visited local municipalities to check on ambient air quality conditions. Traditionally, this had been the responsibility of local environmental protection bureaus, which rarely had sufficient staff or monitoring equipment.

Issues and Analysis

As a result of the review, several issues were identified that require further consideration and analysis. Most of these are related to the application and interpretation of environmental standards and their impact on costs and long-term environmental management of the power plant sector.

Sulfur Dioxide Emissions

The mass-based standard for SO_2 emissions could be interpreted as being overly strict, compared to the concentration-based standard, especially for plants that plan to expand. For example, approximately 0.2 tons per megawatt daily (t/MW/day) equals 8.3 g/kWh of SO_2. According to 1997 data from SPC, the average level of SO_2 emissions from all thermal power plants in China is 84.6 g/kWh, 10 times the World Bank standard. The average level of mass-based emissions of SO_2 from World Bank-financed power plants is 3.69 g/kWh, only 4 percent of the national average. These figures show the positive impact of World Bank-financed plants on the power plant sector.

If the sulfur content of coal could be reduced at reasonable cost (which is questionable) to less than 0.5 percent through coal washing, SO_2 emissions could meet the new World Bank standard (1,200 mg/Nm^3) without flue-gas desulfurization. The sulfur content of coal used by most of the plants in the survey ranges from 0.4 to 0.9 percent.

The heating value of coal and overall plant efficiency are clearly linked. Cleaner coal with concomitant higher heating value means that less coal has to be burned to produce the same amount of energy, which results in fewer emissions, less wear and tear on plant systems, etc. Plants that burn low-quality coal with low heating value and relatively high sulfur content (mostly near mine mouths [e.g., Yanshi]) have difficulty meeting new World Bank standards for emissions.

Figure 2 shows the general compliance of the six plants visited by the review team. All six clearly met the World Bank standard for SO_2 emissions.

Particulate Emissions

The Chinese standard for particulate emissions is much less stringent than the World Bank standard. Nevertheless, three of the six plants we visited, Beilungang I and II, Wujing, and Yangzhou, currently meet the World Bank's

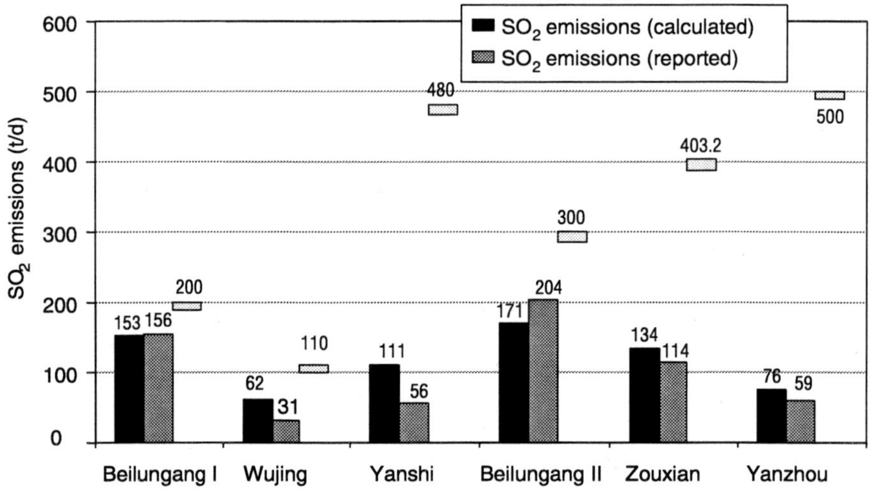

FIGURE 2 Compliance for SO_2 emissions.

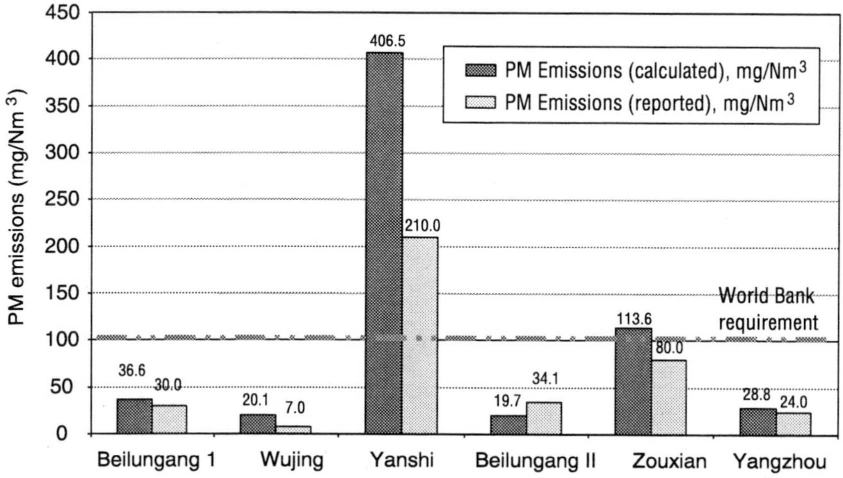

FIGURE 3 Compliance for particulate emissions.

new standard. Two other plants, Yanshi and Zouxian, will have to improve the quality of coal or the efficiency of the electrostatic precipitators to meet the World Bank standard (Figure 3). When the Yanshi plant was built, a low-efficiency precipitator was specified to favor a local contractor, which resulted in the plant being out of compliance.

Options for meeting the new standard should not be limited to improving the efficiency of the precipitator; improving the quality of coal by reducing ash content and increasing heating value should also be considered. According to data provided by Yanshi, the estimated cost of renovating the precipitator is \$2.2 million, equal to \$0.26 million per year for a 20 year loan at 10 percent interest (\$0.14 per ton). This option appears to be much less costly than coal washing, which would cost \$15 to \$30 per ton.

The World Bank standard for particulate emissions was changed in 1997 from 150 mg/Nm^3 to 50 mg/Nm^3. The impact of this change on PM_{10} and smaller particulates[1] (the most dangerous particulates to human health) is still unclear. Even if particulate emissions can be reduced to 50 mg/Nm^3 by improving the efficiency of the precipitator, fine particulate emissions may not be reduced proportionally. Thus, adopting the new World Bank standard may not lead to the expected reductions in PM_{10}. Both the World Bank and the Chinese government are studying a separate requirement for very fine particulate emissions, but neither has adopted one as yet.

[1]The World Bank standard in 1997 was for PM_{10}, but current analyses focus mostly on $PM_{2.5}$.

Modeling of Dispersion and Thermal Pollution

The incremental contribution of power plant emissions to ambient air quality is normally determined by standard mathematical models (usually some variant of a Gaussian dispersion model). The results can then be combined with estimates of background air quality to produce an overall estimate of air quality with the proposed power plant in operation. These estimates can then be compared to air quality standards to determine whether or not the standards would be met.

The environmental assessment process includes a formal requirement for air quality modeling, which is generally done by a local or regional design institute that specializes in assessment techniques. However, after reviewing the dispersion models, it was apparent that the models are not being taken very seriously. The crews that take ambient air quality readings do not compare them against model results, and the coefficients in the models are taken from textbooks. Improving the models will require correlating model results with empirical results.

To get a more complete picture of ground-level conditions based on the concentration of SO_2 in flue gas, ambient air quality at ground level can be determined by calculating the portion of SO_2 that touches the ground under the most unfavorable conditions. In the plants visited by the review team, the maximum additional pollution from any one plant was only about 5 percent. Even though the models on which this estimate is based may be suspect, we can conclude that the portion of urban air pollution attributable to the power sector is minor compared to the portion from other sources.

Generally, the World Bank requires one year of data on ambient air quality as background, and collecting these data may entail significant costs. In addition, between the time background data are collected and the time the power plant goes into operation, ambient conditions may change appreciably, making the impact analysis difficult, if not impossible, to verify. We found this problem at all of the plants we visited. Nevertheless, because modeling indicated that power plant contributions to overall ambient pollution were never more than 5 percent of ambient air quality limits, we concluded that for all practical purposes, power plant emissions superimposed on ambient conditions were well within the standard signal-to-noise ratio for ambient conditions.

Thermal pollution was also the subject of extensive physical and mathematical modeling (e.g., Wai Gao Qiao plant in Shanghai). Thermal pollution modeling, which is normally done when there are cooling system discharges, can be complex and expensive to perform and verify. In addition, thermal pollution is very difficult to measure in situ, and the World Bank standard is somewhat vague. (The 1997 guidelines allow an increase of 3°C at the edge of the mixing zone for waters at 28°C or higher.)

Predictions of ambient air quality and (in some instances) thermal pollution, which are integral to the environmental assessment process, must be verified after plant operations begin to confirm the modeling and check impacts. A review of data from the six plants we visited indicated a lack of commitment to confirming air and water quality predictions, even though the plants were basically in compliance with World Bank and Chinese standards.

Management of Ash

Ash disposal remains a prominent issue because local coals often have a high ash content. Environmental assessments usually include references to using ash as a construction material, but the team did not see efforts being made to do this. Nevertheless, ash was managed in a relatively sound and environmentally safe way at all of the plants we visited. Ash is disposed of in large ash ponds (measuring several hundred meters on each side), a method that requires storage silos at each plant, and transport either via sluice pipes several kilometers long, truck, or rail. The facilities for treating ash add significantly to the cost of operations. Disposal sites tend to be near bodies of water (mud flats) or in areas where farming is a marginal activity. Ash yards require significant civil works to ensure that the materials are contained and that drainage is strictly controlled. At some plants, including the most modern plant we visited, the ash yards and ponds were nearing capacity, and new sites for ash disposal will be required.

In some cases, ponds have become desiccated, and ash has become airborne as particulate, which has caused problems for farmers and residents. In other areas, rain has washed the ash off site. Because storage capacity near the plants is limited, more mine-mouth methods, such as reducing ash via coal washing, should be explored.

Water Pollution

To meet World Bank and Chinese standards, municipal wastewater from power plants requires secondary treatment prior to discharge. At some plants, the volume of wastewater was too small to justify a wastewater treatment system. At other plants, the wastewater treatment process was the last system to be implemented. Because wastewater usually includes cooling water and sanitary water and, sometimes, coal pile runoff, it is difficult to treat via biological processes. Nevertheless, typical, off-the-shelf biological processes are being used, and not always effectively. In some cases, the influent water for cooling is of such poor quality that the effluent standards actually represent an improvement in water quality. Based on this finding, some have concluded that power plants should not be required to treat effluents at all.

CONCLUSIONS

China has a number of world-class power plants equipped with the latest pollution abatement technology and has made significant progress in developing its power industry. In addition, China has taken the first steps toward controlling pollution from that industry. Nevertheless, compliance monitoring and enforcement are not uniform and are not implemented everywhere in the country. Significant progress has been made in assessing the environmental impact of proposed plants, but not as much progress has been made in ensuring that pollution control systems are operated once the plant is built. As China becomes more prosperous, the need for electrical energy will increase; hence it will burn more coal. It is, therefore, imperative that power sector pollution control remain a high priority.

BIBLIOGRAPHY AND REFERENCES

Jia, L., Baratz, B., and Fritz, J. 2000. Environmental Performance of Bank-Financed, Coal-Fired Power Plants in China. Washington, D.C.: World Bank, East Asia Environment and Social Development Unit.

World Bank. 1998. Pollution Prevention and Abatement Handbook. Washington, D.C.: World Bank.

World Bank. 2001. China: Air, Land, and Water Environmental Priorities for a New Millennium. Washington, D.C.: World Bank.

Prospects for Distributed Combined Cooling, Heating, and Power Systems in China

LIWEN FENG
Falcon Group Inc.

YINGSHI WANG
Institute of Engineering Thermophysics
Chinese Academy of Sciences

Distributed energy systems can complement large thermal power plants and large steam generators. A distributed-energy power-generating system is generally modular, can provide anywhere from several thousand kilowatts (kW) to 50 megawatts (MW) of power, and is located near its customers. It requires almost no public power or steam network, and, because it is located near customers, transmission losses are significantly reduced.

The concept of distributed energy was initiated and promoted in the United States and was subsequently accepted by other industrialized countries. Active research in the technology for distributed energy began after the energy crisis of 1973. In recent years, as a result of the serious ecological and environmental challenges facing traditional power sources (e.g., global warming), the pace of the development of conventional plants in industrialized countries has slowed. At the same time, the development of nuclear power is being restricted because of safety concerns. For all of these reasons, developed countries are now actively pursuing research and development of distributed energy technology.

Although distributed energy technology has been researched for decades, rapid development has occurred only in the past few years. Many experts and scholars now consider distributed power to be the "energy for the 21st century." As energy infrastructure is restructured and new technologies are developed, and as the demand for high-quality energy increases, distributed energy will become increasingly attractive. Distributed energy systems are environmentally friendly, highly efficient, and highly flexible, and they can be combined with the traditional energy grid to improve energy services for all people.

The power-generating equipment in a distributed system can be a small (micro) gas turbine, a heat pump, or a fuel cell. Energy can be produced from solar (electric or thermal), geothermal, wind, natural gas, and methane resources. Second-generation distributed energy systems are characterized by five features: many possible fuel sources; small or micro equipment; combined cooling, heating, and power (CCHP) generation; intelligent management; and a high standard of environmental protection. Distributed energy systems are not likely to replace traditional energy systems. However, when combined with traditional systems, they can make overall systems more efficient and more robust.

ADVANTAGES OF DISTRIBUTED ENERGY

Although distributed energy technologies are still in their infancy, they have numerous advantages over traditional energy technologies: reduced emissions; increased efficiency; flexibility; improved safety; and load balance.

Lower Emissions

Because distributed energy systems use renewable energy sources and gas turbine engines with extremely efficient combustion chamber technology, emissions of nitrogen oxides (NO_x) can be reduced to less than 25 parts per million (ppm). Emissions of carbon dioxide (CO_2), sulfide, and dust can also be reduced significantly compared with emissions from large thermal power plants.

Efficient, Comprehensive Energy Consumption

CCHP systems are small scale and flexible. These systems can effectively integrate electricity, heating, and cooling requirements, supply enough energy to meet demand without waste, and eliminate the need to transmit cooling and heating energy over long distances. The efficiency of a distributed energy system can be higher than 75 percent.

Intelligence and Flexibility

Because the overall capacity of a CCHP system is relatively small, and because start-up, shutdown, and regulatory requirements can be met quickly, a distributed system can operate automatically, without personnel on duty, and operation is flexible and easy.

Safety

As recent power blackouts have shown, the traditional electricity grid is still heavily dependent on a stable exterior environment, and the political and economical effects of damage to transmission lines are becoming more and more serious. However, if a large grid is combined with distributed energy located near its customers, it can maintain the power supply even if the grid collapses or there is an accident, such as an earthquake, snow storm, or even a war. In addition to standard power, distributed energy can supply emergency standby power and, if necessary, extra power during peak times, thus greatly improving the reliability of the power supply.

Supply Balance

To reduce adverse effects on the environment, natural gas is replacing coal. But with the increased use of air conditioning, imbalances in the electricity load between winter and summer are becoming common in large cities, including Beijing. The supply of natural gas is high in winter and low in summer. Distributed energy systems can balance the power supply to some extent, because they use natural gas in summer to generate power for air conditioning. Thus they improve the utilization rate of grid equipment and natural gas pipelines.

THE INEVITABILITY OF DISTRIBUTED ENERGY SYSTEMS IN CHINA

China is developing rapidly; however, the relative shortage of energy resources and the waste caused by the inefficient use of existing resources could keep China from sustaining that rate of development. In terms of per capita resources, China has a serious shortfall in primary energy. China has approximately 21 percent of the world's population, but only 11 percent of the world's known coal reserves. Global coal reserves are expected to last for 230 years, but China's reserves are expected to last for only 90 years. China's reserves of crude oil account for only 2.4 percent of total global reserves. Global crude oil reserves are expected to last for 48 years, but according to the Chinese Statistical Bureau, China's reserves will last for only 22 years. Therefore, more complete combustion and more efficient energy use will be absolutely essential to continued modernization in China.

The emission of greenhouse gases is a particularly serious problem, and China ranks second in the world in greenhouse gas emissions. The Chinese government has decided to participate in the Kyoto Agreement and to undertake the obligations stipulated therein, which means that from 2008 to 2012, China has agreed to reduce greenhouse gas emissions to the level of 1990. To realize this goal, the use of coal will have to be considerably reduced and the development of, and exploration for, natural gas will have to be considerably increased. The development and use of solar, wind, and other renewable energies are becoming inevitable.

The use of natural gas will become a significant issue for the Chinese energy industry in the next 10 years. But simply replacing coal with natural gas for power and heating will not reduce greenhouse gas emissions significantly. In addition, it will increase fuel prices significantly and cause a huge loss of revenue to the state and private industry.

Economic development calls for a safe, stable power system. Recent blackouts in the northeastern United States, Canada, London, and Taiwan should remind us to take the issue seriously. As China's economy develops rapidly, the centralized power supply grid, along with its inevitable safety problems, will also expand rapidly. Reasonable development of a safer power supply structure would combine distributed power with the centralized power supply.

DEVELOPMENT OF DISTRIBUTED ENERGY

Table 1 shows the status of distributed energy projects in China in 2003. To meet environmental protection standards, remain competitive, and incorporate technological advances, a long-term energy plan must be both flexible and practical. The plan should include a reasonable layout of the large grid and the development of distributed energy.

TABLE 1 Status of Current Distributed Energy Projects in China

Project Location	Equipment Status	Remarks
Shanghai Huangpu Center	1 × 1000 kW Solar Saturn 20 diesel gas turbine; 1 × 3.5 t/h HRSG	Operating
Shanghai Pudong Airport	1 × 4000 kW Solar natural gas turbine	Operating
Shanghai Minhang Hospital	1 × 400 kW Jiantai gas engine; 1 x 350 kg/h HRSG	Operating
Guangdong Dongguan Shoe Plant	11 × 1020 kW diesel gas engine; 11 x 0.5 t/h steam boilers	Operating
Guangzhou Aluminum Group	1 × 725 kW heavy oil gas engine; 1 × BZ200 waste heat direct engine	Operating
Shanghai Engineering University	1× 60 kW Capstone micro-engine; 1 × 150,000 Mega Pascal waste heat direct engine	Under development
Beijing Gas Group	1 × 480 kW, 1 × 725 kW gas engine	Installed
Beijing Ci Chumen Station	1 × 80 kW Bowman gas micro-engine	Installed
Beijing International Trading Building	1 × 4000 kW Solar Centaur 40 gas turbine	Under development
Beijing International Shopping Mall	1 × 4000 kW Solar Centaur 40 gas turbine	Under development
Soft Ware Square	1 set of 1200 kW Solar gas turbine	Tendered

- For the remote, less-developed areas of western China, distributed energy systems can provide strong support during the period of economic development in a short period of time and at a relatively small cost because they can make use of the rich natural resources of these areas and diverse forms of renewable energy.
- For the economically developed southeastern coastal areas, the requirements for environmental protection are becoming more stringent, and demands for energy products, such as cooling, heating, and electricity, are becoming more diversified. The development of distributed energy along with traditional power generating systems will improve the comprehensive use of energy sources, reduce energy supply costs, and improve the safety of the power supply to urban areas. At the same time, a distributed energy system will improve environmental protection. Distributed energy is the optimum choice.

POWER GENERATION BASED ON NATURAL GAS

Known global natural gas reserves exceed 140 trillion cubic meters (m^3) and are expected to last for about 68 years if they are consumed at a rate of 2 trillion m^3 annually. The exploration and use of natural gas in China are relatively underdeveloped; the known reserve is only 1.2 percent of the global reserve, and the current annual output is about 22.5 billion m^3 per year. Geologically, the natural gas reserves in China are estimated at more than 30.8 trillion m^3. According to international general practice, about 1.02 trillion m^3 can be excavated for 45 years, and improvements in transporting gas from west to east in China will lead to an unprecedented change in the Chinese energy structure.

Natural gas will be used primarily to alleviate serious pollution in cities. There will, however, be negative consequences if natural gas is not used appropriately. China must improve the utilization rate of energy as it addresses environmental contamination. Distributed energy can provide a new, realistic method of using natural gas that will increase energy efficiency, offer economic benefits, and improve safety. An example of the use of distributed natural gas power is shown in Figure 1.

DISTRIBUTED ENERGY SYSTEMS BASED ON RENEWABLE ENERGY

According to documents provided by the expert committee of the "863 Program," China is rich in renewable energies—both in quality and quantity. For example, the annual radiation of solar energy for more than two-thirds of China exceeds 600 MJ/cm^2; the annual solar energy radiation absorbed by the Earth's surface is equivalent to 17 trillion tons of standard coal. The prospective reserve of geothermal energy in China is about 135 billion tons of standard coal (the explored coal reserve is about 3.16 billion tons).

FIGURE 1 The Beijing Gas Group Office Building (23,000 m^2) is already built and is being commissioned. Major energy equipment includes: 1 set of 480 kW + 1 set of 72 kW Caterpillar gas engines; and 1 set of BZ100 + 1 set of BZ200 waste-heat direct-absorption gas turbines. The system guarantees basic power needs. Any shortfall for heating, cooling, or hot water is supplied by the grid.

The main obstacles to the use of renewable energies are low efficiency, low density, and instability of the power supply; all of these make it difficult to integrate renewable energies into the centralized power supply. By integrating with distributed energy, renewable energy can significantly improve the efficiency of the power supply on the basis of cascade utilization. The energy density for distributed energy is far lower than for the centralized power supply. Moreover, by using modern, energy-saving technologies, distributed energy can overcome the instability of renewable energy supplies to a great extent. In fact, distributed energy has provided a new impetus for the development and use of renewable energy and can significantly improve the efficiency of the power supply, provide economic benefits, and improve safety.

PROBLEMS ENCOUNTERED

Problems encountered in the promotion of natural gas to supply power include: interconnection; sales; fire regulations; pricing; and project operation and taxes.

Interconnection

CCHP systems can produce three types of energy—cooling, heating, and electricity. Cooling and heating demands can be met directly, but the electricity supply must be interconnected with the city grid for economic and safety reasons. At present, the electric power authorities in Beijing are resistant to the idea of interconnecting the main power system to other power generating sources, because the department believes that the interconnection of small power sources in large quantity will affect the operational safety of the grid. However, small generators that satisfy national standards are mature enough for interconnection with the grid. There may be some problems in terms of dispatching and managing energy, but, extrapolating from broad experience, these difficulties can be overcome. In addition, there are some examples of successful interconnections, such as Shanghai Pudong Airport and Minhang Hospital.

The power department in Beijing has invested economically in most power supply facilities. When customers build their own CCHP systems, their power generating costs will be lower than the cost of power generated by the city grid, and customers will naturally use the lower cost power, which will negatively affect the power department's revenues. This issue can be resolved by compensating the large grid with a portion of the revenue received by the owners of small power sources.

Chinese laws stipulate clear rules for the interconnection and construction of power sources. If we combine foreign experience with these standards, we can formulate guidelines for interconnecting small power sources that are suitable to the Chinese situation and take into account the interests of the power department and the technical issues surrounding interconnection.

Sales

Chinese laws stipulate that unless the power supply enterprises agree, no other entity is allowed to supply power. This inhibits the range and economics of distributed power sources. For example, if several different entities wish to invest jointly in a small power supply center to save money and improve the energy utilization rate, under the current rules they cannot. Thus, current regulations inhibit the development of small power suppliers.

In addition, the current grid system charges one tariff for all power, regardless of how the power is generated. This is unfair to both gas cogeneration and small power suppliers because the price of natural gas is much higher than the cost of coal, and therefore the cost of gas generation is much higher than the cost of coal generation. However, coal generation does not have the environmental benefits of gas generation. To address these issues fairly, the government could subsidize gas generation.

Fire Regulations

Current fire regulations in China stipulate that the pressure of gas entering each home shall be at moderate levels. However, gas turbines generally require much higher pressure. Thus, current rules inhibit the development of gas-generated CCHP systems in cities. In Japan, the United States, and other countries, the pressure of the gas entering each home can be higher than in China. Instead of limiting the use of high-pressure gas in those countries, management and technical measures have been developed to solve possible problems arising from high-pressure gas.

Energy Pricing

The high price of gas in China has been the main factor inhibiting the development of gas generation. The price of natural gas in China is double the price in Russia and the United States. Aside from its environmental benefits, gas-generated CCHP systems would increase the quantity of gas used in Beijing and balance gas utilization in winter and summer. Therefore, the government and Beijing Gas Group should offer policies that encourage the use of gas-generated CCHP systems.

Project Operation and Taxation

The state encourages the development of gas generation and has listed co-generation in its catalogue of "Industries, Products, and Technologies."

In general, gas generation is an energy-saving, environmentally beneficial tool that uses new technologies. But if all of the economical and technical risks of such projects must be borne by customers, the projects will certainly be negatively affected. For this reason, the economic and trading commission has introduced energy management company (EMC) energy-saving operating methodologies from abroad.

The World Bank offers loans for Chinese companies that use EMCs. In the initial loan phase, these companies enjoy reductions in and exemptions from import duties, as well as favorable polices regarding the value-added tax on imported equipment. However, companies using EMCs are no longer allowed to receive World Bank loans directly, and so are required to pay the regular customs duty and value-added tax for imported equipment. Because gas-generation technology is relatively new, most of the equipment is imported. For example, Zhong Guancun Software Park requires about 10 million yuan to import a 1,200 kW gas turbine, of which about 2.3 million yuan is for customs duty and value-added tax. This significantly increases the cost of the project.

ASSESSING PROJECTS

Difficulties have been encountered in promoting the use of gas-fueled CCHP systems. Because gas generation is not only beneficial to the environment, but would also save energy and balance the peaks and valleys of the gas and power supply in Beijing, it is essential that we try to elicit more favorable policies from the government.

Although distributed energy systems have a number of advantages and are relatively well established internationally, these technologies are still in the beginning stages of use in China. In fact, only CCHP is mature and can be put into market operation. CCHP is being used in the newly built Pudong Airport in Shanghai and at Minhang Hospital in Xinzhuang Shanghai. In Beijing, CCHP projects include Beijing Gas Monitor and Control Center, which uses a gas engine, and Ciqumen Station, which uses a micro gas turbine.

The key to the success of CCHP systems, illustrated by examples in both China and abroad, is selecting equipment and formulating a plan of operation that satisfies customers' requirements for a sufficient supply of natural gas. CCHP projects must therefore be based upon system capacity and local energy pricing. We have already investigated several projects. For example, the owner of a supermarket in Wuhan, in Hubei Province, who is very interested in CCHP, hired us to develop a plan. However, after familiarizing ourselves with energy prices in Wuhan, we found that the investment would not be returned and that CCHP is not suitable for the area.

Once a project is verified as feasible, the issues of load analysis and system capacity must be faced. Based on our experience, this step is critical to the project's success. An example of a failed project was a CCHP system for a hospital in Shanghai, which failed because the unit capacity far exceeded the actual load of the project. Load analysis is as basic to each project as laying the foundation of a house. Analysis involves collecting statistics of the annual operation of office buildings, hotels, and apartments in Beijing, sorting the relevant cogeneration load rule, and making load forecasts as close as possible to the actual operation of the conditions of potential customers.

Selecting the unit is also essential. There are many differences between a CCHP project and a traditional plan. First, the energy supply source is relatively independent, which means that it does not depend on the larger grid network in most cases. It is not economical to purchase all the power from the city grid, so there is a difference between the unit capacity of the CCHP system electrical load and the unit capacity of economical operation. If the guiding principle of "the quantity of interconnected power depends on the quantity of supplied cool and thermal, interconnected but not on grid, and the shortage of power shall be supplemented by the power of city grid" is applied in a simple way, the CCHP unit selected will produce a scenario in which a small cart is pulled by a large horse, which will not yield an economic benefit. Therefore, the economic operation load, rather than the peak load, should be the reference point in selecting a CCHP unit.

There are various kinds of generators for CCHP projects: gas turbine, gas engine, gas exterior engine, and micro gas turbine. There are mainly two types of waste-heat recovery units: waste-heat recovery boilers and waste-heat, direct-fire absorption engines. Generally speaking, the generation capacity is usually below 1,000 kW, for which a gas engine, micro gas turbine, or gas turbine is most appropriate. Although, the most widely used and most mature product is the waste-heat recovery boiler, the waste-heat direct absorption engine is a focal point of CCHP research. With the support of the U.S. Department of Energy, the state of Texas plans to build a distributed CCHP (DCCHP) project that uses a waste-heat, direct-absorption engine. The Texas DCCHP project will be the largest in the world; it will involve seven companies, including the Yuan Da Group from China. However, the generator and waste-heat recovery method are still to be decided based on comparisons of several plans.

A CCHP system is different from the single operation of a generator, as well as from the traditional cogeneration power plant. In most cases, cooling and heating are supplied independently, and the load is rarely stable. Each season of the year has a unique power load. Weekend and weekday loads differ, as do day and night loads. Because the system is not buffered by support from the city grid, it has to follow changes in customer load closely to adjust the supply of cooling and heating to maximize the utilization rate of the energies. A CCHP system does not require manpower on duty.

SUPPORTING THE DEVELOPMENT OF DISTRIBUTED ENERGY

The main obstacle to the development of distributed energy in China is not the lack of technology or capital, but the rules and policies of the state that are at odds with the effective use of technology and the lack of an effective supervisory mechanism. The measures suggested below are critical to the development of distributed energy in China. In all cases, the power grid must provide standby power security to customers who use distributed energy facilities.

First, when customers who use distributed energy facilities require emergency power, the tariff may be higher than for other customers. This capacity tariff, or "power source standby fee," should not be allowed because the capacity for distributed energy customers is so small that the power grid does not have to maintain the equivalent standby capacity.

Second, distributed energy facilities may sell extra power to the grid, and the grid may be obliged to off-take such power. In that case, the tariff should be 90 percent of the tariff paid by customers; this will reduce transmission loss and wasted energy.

Third, distributed energy facilities should be permitted to interconnect to the main power system. To ensure the safe use of power by customers, the distributed power should be connected to the power supply terminal or customer terminal of the customer transformers. Because the capacities of distributed facilities are not sufficient to affect the safety of the grid, the grid should be able to address this issue if safety of the system is affected.

Fourth, the distribution of natural-gas-generated energy should be optimized to reduce adverse environmental impacts. This optimization should be a priority, regardless of industry barriers, and should provide natural gas to the places most in need to reduce the price of energy. The use of natural gas by power enterprises should be strictly minimized. CCHP systems should be used to satisfy peak demand for power and balance the cost during off-peak times to improve the peaking capability of the power plant.

DISTRIBUTED ENERGY SYSTEMS AND THE BEIJING OLYMPICS

Energy has long been closely related to the quality of life and has played an important role in large gatherings of people. The Beijing Olympics in 2008, which will be viewed by the whole world, will be the largest gathering ever held in China. The world has high expectations for the Beijing Olympics, and this is an opportunity for China take pride in its accomplishments. The Olympics will be a large, complicated project, and energy construction will be an important part of it. The "Energy Construction and Structure Adjustment of Beijing Olympics Action Plan" proposes using distributed energy, including thermal pumps, photovoltaics, solar thermal energy, and wind technologies. This is the first time new energy would be used for the Olympics.

"Green Olympics, Scientific Olympics" is China's commitment to the world and to ourselves. Criteria for defining "green" include how energy is produced and how it is used. For a long time, we have automatically relied on electricity and heat supplied by the city grid and networks. However, a key issue for China is avoiding, or at lease reducing, electricity loss resulting from long-distance transmission and heat loss from long-distance pipelines. Another is assessing environmental contamination caused by emissions from large thermal power plants. Of course, electricity supplied by city grids has advantages—it is stable and reliable, and unit costs are low. But as new energy technologies are developed and paired with distributed energy, these advantages will be less striking.

The Beijing municipal government and the Beijing Olympic Committee have supported the building of a green energy center near the Olympic Park that will use various kinds of distributed energy technologies. A CCHP network will be built near the park, and various kinds of energy facilities will be connected with this network from different locations. Those facilities will include natural gas heating, electricity, and cooling, fuel cells, solar electric and thermal energy, and wind energy. A control center will be built to monitor and supervise the operation of each type of facility, load requirements, energy supplies, and the use of renewable and green energies to optimize their comprehensive benefits.

SUMMARY

Distributed energy is only one direction in energy technology development, but it is especially important to the current stage of energy restructuring in China. Although our economy is accelerating rapidly, many aspects of it are still in the transition stage, and the energy industry is adapting to economic developments. Distributed energy development not only embodies technological progress but also represents a transformation in the way we think about energy use.

Of course, there are still many difficulties to overcome in the development of distributed energy in China. Some of those difficulties are related to policies and technologies, but most of them are related to interconnection, energy prices, and supporting policies. Despite these challenges, distributed energy technology development is an inevitable trend that will help meet the needs of our country's economic future together with traditional energy methods.

BIBLIOGRAPHY

Beijing Natural Gas Utilization Research Institute. 2003–2004. BCCHP System vs. Natural Gas Utilization: Research and Pilot Project. Beijing: Natural Gas Utilization Research Institute.

Chen, H. 2001. Improving Efficiency to Support Sustainable Utilization of Energy. Beijing: Department of CHP and Energy Saving, State Planning Commission of China.

China Energy Conservation Corp and China SDPC. 2003. China Micro (Small) CCHP Technology Potential Market Analysis. Beijing: SDPC.

Chinese Statistical Bureau. Multiple years. Chinese State Year Books. Beijing: China Statistical Publishing House.

IEA (International Energy Agency). 2003. World Energy Investment Outlook. Paris: IEA.

IEA, in cooperation with the Energy Research Institute of the China State Development Planning Commission (SDPC). 2002. China Energy Outlook. Paris: IEA.

Wang, J. 2002. China Energy Development Report. Beijing: Beijing Market Economic Development Research Institute.

Power-Sector Energy Consumption and Pollution Control in China

XUCHANG XU, CHANGHE CHEN, HAIYIN QI, DINGKAI LI, CHANGFU YOU, and GUANGMING XIANG
Tsinghua University

Energy consumption and urban air pollution have been inextricably linked in China for decades, because coal combustion is the main source of fuel for the Chinese economy. Because of growing pollution problems, concerns about global warming, and expected increases in private transportation, this a critical time for Chinese energy policy makers. In addition, energy policy in China has far-ranging effects on the rest of the world because of China's increasing demand for oil and gas, as well as other commodities, such as steel. Thus everyone has a stake in seeing China develop a sound and sustainable energy policy. Fortunately, the Chinese government, which manages the economy as a whole, is in a position to make this happen.

In 2000, total energy consumption in China was 1,357 million tons carbon equivalent (Mtce) (i.e., 950 million tons oil equivalent [Mtoe]). As Figure 1 shows, total energy consumption in China is less than consumption in the United States but greater than consumption in India. The gross domestic product (GDP) of China was also much smaller (only about one-fifth) than the GDP of the United States. Figure 2 shows the average per capita energy consumption for China and for other countries in kilograms coal equivalent [kgce]. In 1990, per capita energy consumption in China was 1,110 kgce, or 1.11 tons carbon equivalent (tce), per capita (i.e., one-tenth the per capita consumption in the United States). By 2000, that figure had increased to 1.35 tce per capita (i.e., one-seventh that of the United States). By 2010, the number is expected to reach 1.6 to 1.85 tce per capita. Although China's per capita energy consumption is similar to that of many other developing countries, because of the size of the population,

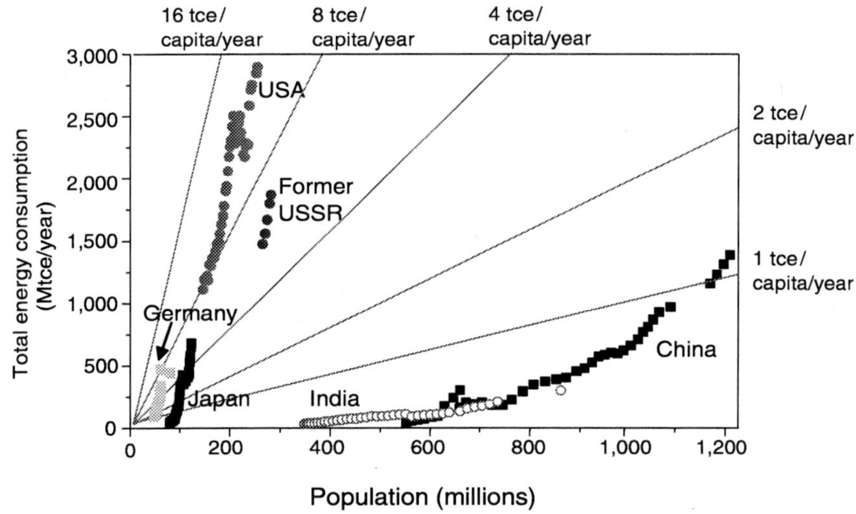

FIGURE 1 Energy consumption by population for different countries.

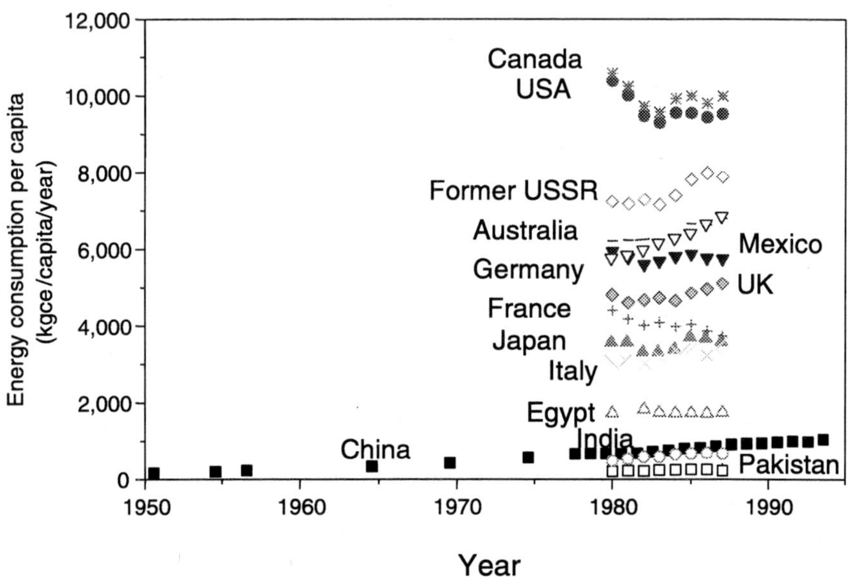

FIGURE 2 Average energy consumption per capita for different countries (in kilograms coal equivalent [kgce; 1,000 kgce = 1 tce]).

China is one of the largest consumers of energy in the world. Therefore, energy consumption and pollution must be carefully regulated.

By 2050, the International Energy Agency (IEA) estimates that total energy consumption in China could reach 3,440 Mtce, 2.5 times the consumption in 2000. This estimate is in agreement with an assessment by the Chinese Academy of Engineering published in 1998 (see Figure 3). If current trends continue, the population of China in 2050 could reach 1.6 billion, and average energy consumption per capita could reach 2.15 tce (i.e., one-quarter of energy consumption in the United States in 2000). At that point, China would be tenth in the world in terms of per capita energy consumption.

For the next 50 years, perhaps even 100 years, most of the energy consumed in China will continue to come from coal (Figure 4). Currently, 70 percent of total energy in China is produced from coal. The percentage is expected to drop to

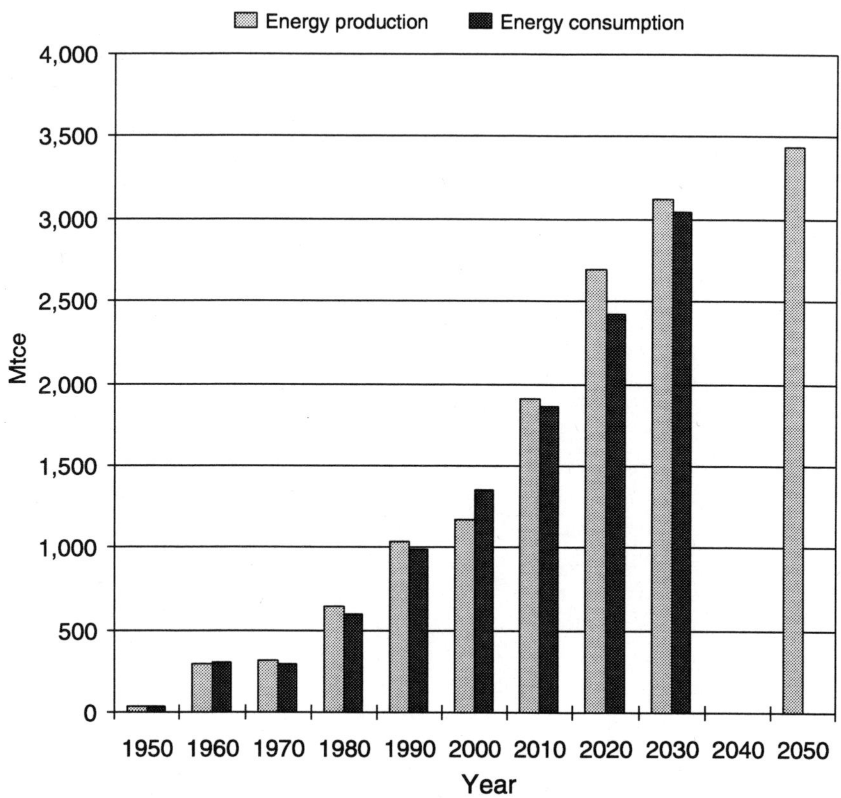

FIGURE 3 Primary energy production and consumption in China.

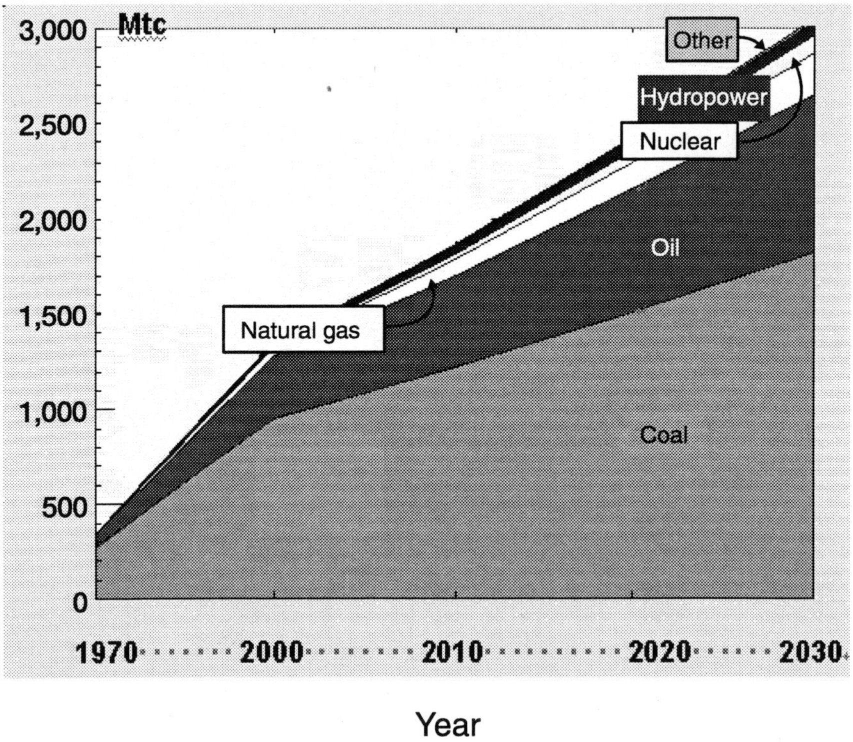

FIGURE 4 Estimated energy consumption in China, 2004–2030.

66 percent by 2020, but the total amount of coal consumed is expected to increase to 1.2 to 3.1 billion tons, or even 4 billion tons, depending on the level of economic development and the cost of other fuels. In any case, China must focus on reducing air pollution from coal combustion.

Coal is the source of a moderate fraction of the energy consumed in most developed countries. For example, in 1999 in the United States, only 25 percent of total energy came from coal. The percentages for Japan, France, and several other developed countries are similar. China, however, will continue to rely heavily on coal because it is the largest locally exploitable fossil resource. In fact, China has few other choices at this time. Therefore, China must balance the use of coal against environmental protection, which will require reducing overall energy consumption and developing and implementing pollution-control systems.

THE POWER SECTOR

The total electric power capacity in China in 2000 was 319 gigawatts-electric (GWe), of which 230 GWe was from coal. It is estimated that the total electric power capacity in China will rise to 580 GWe by 2010, of which 350 GWe will be from coal. Even though China's total electric power generating capacity is the second highest in the world, annual average per capita consumption is very low, close to 1,100 kilowatt-hours (kWh) (i.e., an order of magnitude less than the average consumption level in the United States), as shown in Figure 5. Even in 2050, if the Chinese economy continues to develop, the annual per capita average might be only 4,460 kWh, close to one-third that of the United States in 2000. Because of China's large population, however, which could reach 1.6 billion by 2050, the overall annual power consumption level could be as high as 1,560 GW.

In the United States, almost 91 percent of coal consumption is used for electric power generation. The total amount of coal consumed in the United States is less than in China. Chinese coal-based power generation increased from 25 percent in 1990 to 47.8 percent in 2002; China also uses coal power for industrial and domestic purposes (cooking and heating). Recently, China has made major efforts to change the energy mix to reduce air pollution in key cities; for example, China has tried to import natural gas and to produce natural gas and oil for domestic use. Because the primary energy mix has improved in some large

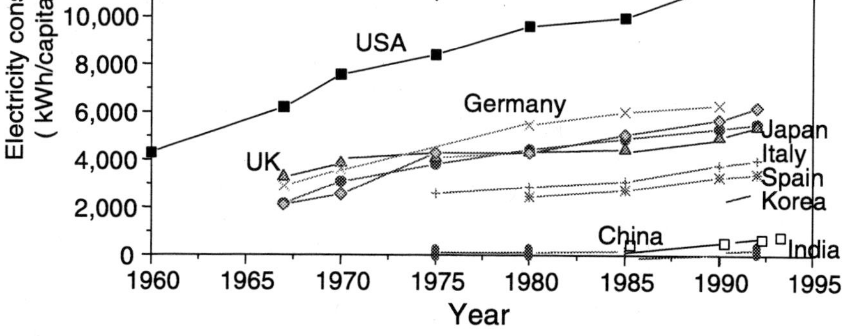

FIGURE 5 Average electricity consumption per capita in different countries.

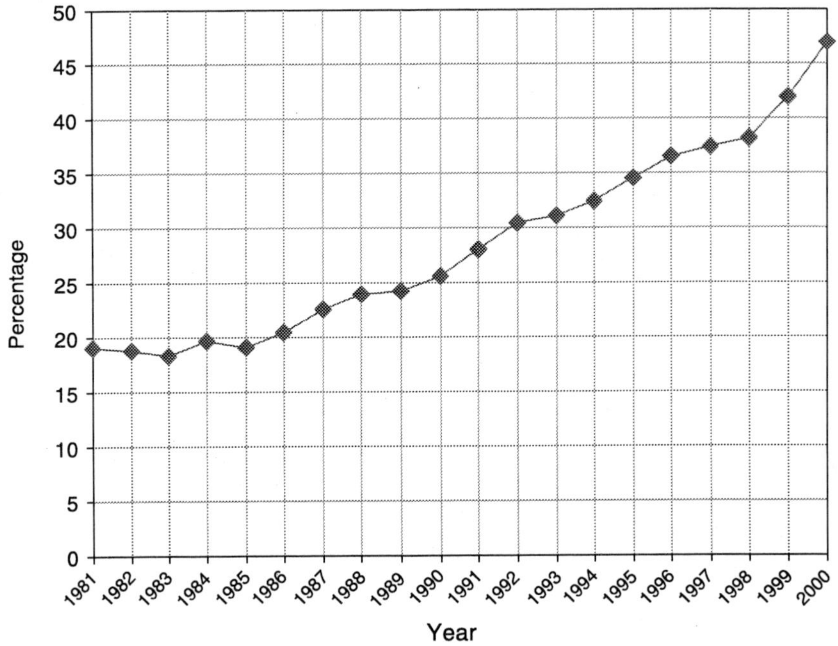

FIGURE 6 Percentage of coal consumption for electricity generation in China.

Chinese cities (e.g., Beijing), the ambient air quality in those cities is now much better than it was 10 years ago.

Because China will have to rely increasingly on imported oil and natural gas imports, the strategy will be to build large, centralized coal-burning facilities to control air pollution. Large power plants can be built or retrofitted with pollution-control technologies, unlike the thousands of small boilers prevalent in Chinese cities. Even though the percentage of coal consumption for electricity generation could therefore continue to increase (Figure 6) and could reach 65 percent by 2050, it would still be lower than the more than 90 percent in the United States in 1999. Therefore, air pollution from coal consumption in the power sector in China is a relatively manageable problem.

CONTROL OF SULFUR DIOXIDE EMISSIONS

China's dependence on coal as an energy source is not expected to change in the near future. Thus, two strategies for environmental control should be considered: (1) improving the efficiency of the energy-conversion process; and (2) improving the efficiency of energy consumption. Tables 1 and 2 show the

TABLE 1 Improvements in Energy Efficiency in China

Energy Consumption by Product or Sector	1980	1985	1990	1995
Electricity (coal) (gce/kWh)	448	431	427	412
Steel (kgce/t)	2040	1746	1611	1519
Cement (kgce/t)	206.5	201.1	185.3	175.3
Synthetic ammonia (kgce/t)	3021	2358	2263	2090
Transportation by train (kgce/10^4tkm)	147.1	118.7	84.2	58.6

TABLE 2 Results of Investments in Energy Efficiency

	1980–1990	1991–1995
Total investment	27.2 billion yuan	40 billion yuan
Energy savings	55.8 Mtce/year	61 Mtce/year

large investments to improve energy efficiency and implement energy-saving technologies in China between 1980 and 1995, and the positive results. These investments are expected to continue. Since 1950, coal consumption per kWh has been reduced by half (Figure 7), but compared with developed countries, it is still high. One way to address this problem is to use supercritical boilers, which are used elsewhere in large coal-fired power plants with thermal efficiencies as high as 41 to 47 percent. Supercritical boilers offer a proven, commercially available way to improve efficiency at a competitive cost relative to other clean-coal technologies

Overall, in 1997, China consumed 8 to 9 percent of the world's energy total but was responsible for 15.1 percent of SO_2 emissions. A year before that, in 1996, total SO_2 emissions in China were 23.46 million tons (Mt)—considered the highest in the world at that time. Since 1996, SO_2 emissions have decreased slightly as coal consumption has been reduced as a result of adjustments in the economic structure of heavy industry (Figure 8). China emits about 10.1 percent of global NO_x emissions and 9.6 percent of global CO_2 emissions.

According to 1997 data, coal combustion accounted for approximately 87 percent of SO_2 emissions, 71 percent of CO_2 emissions, 67 percent of NO_x emissions, and 60 percent of particulate emissions. The huge emissions of SO_2 and NO_x have already caused a great deal of damage to the environment in China. For example, about 40 percent of the total area of China (mostly in eastern China) has pH values of less than 5.6 (see Figures 9 and 10). In the hardest hit areas, pollution has affected both public health and agricultural yields. According to an estimate by the Chinese Research Institute of Environment, the total economic

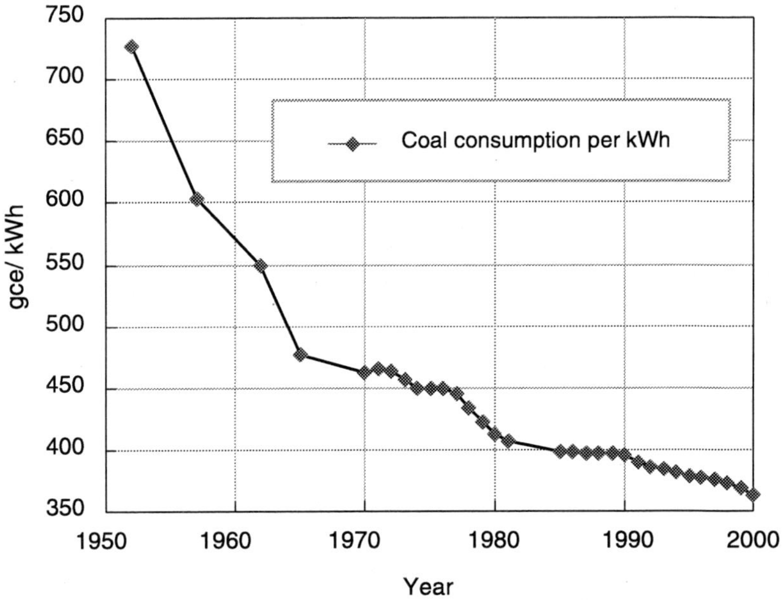

FIGURE 7 Coal consumption for electric power generation (gce/kWh).

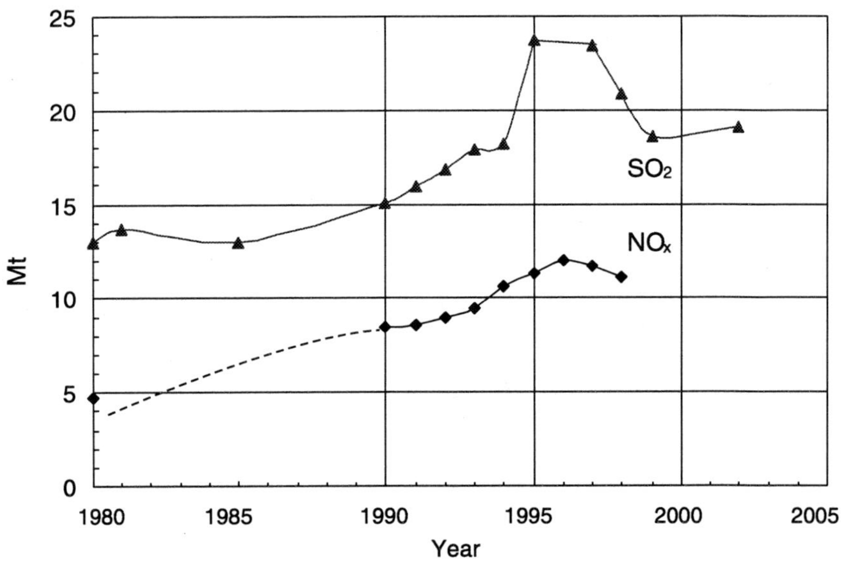

FIGURE 8 SO_2 and NO_x emissions (Mt/year).

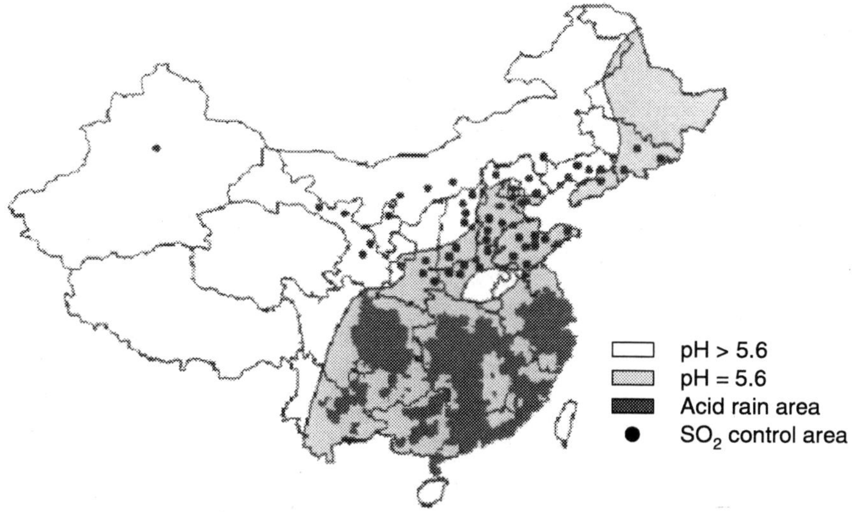

FIGURE 9 Damage from SO_2 and NO_x emissions.

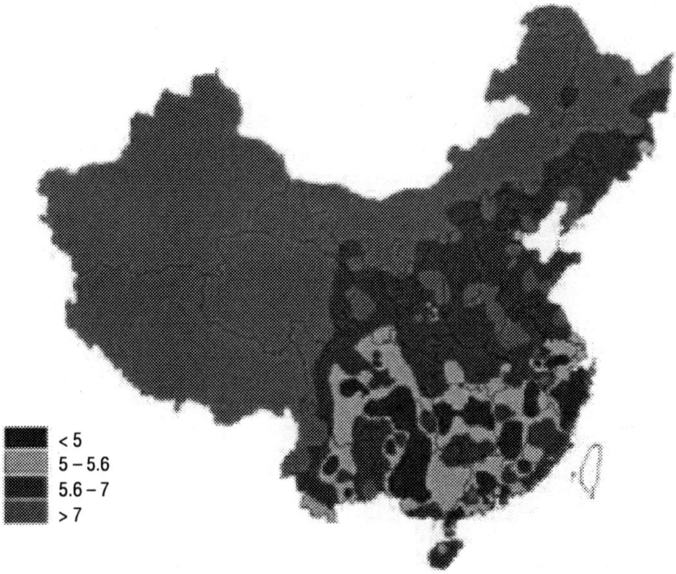

FIGURE 10 Map showing pH values of soil in China.

TABLE 3 Total Economic Loss from Acid Rain and Acid Deposition in China

Economic Loss (billion yuan)	Year	
	1995	2000
Agriculture	21.77	28.48
Forestry	77.58	128.44
Human Health	17.19	19.60
Total	116.54	176.42

loss from acid deposition in China in 1995 was 116.54 billion yuan; in 2000, just five years later, it reached 176.42 billion yuan (Table 3). Pollution from coal combustion has also been an impediment to economic and social development in China.

The United States, the European Union, and Japan have all invested significant resources in fundamental research to control emissions from coal combustion, and many new power-generation techniques using coal could be used in China. For example, combined-cycle power generation and fuel cells (powered by gas made from coal) have a high energy-conversion efficiency and produce low emissions. But even if these technologies are used to produce 5 GWe by 2010, they would account for only 1 percent of the total power generation in China, which will total 580 GWe. Virtually all new coal-fired power plants and some older, retrofitted units use flue-gas desulfurization (FGD) and particulate controls via electrostatic precipitators (ESPs). Nevertheless, pollution control of direct coal combustion must remain a high priority for both the government and the private sector.

The most economical way to reduce SO_2 emissions is to burn low-sulfur coal ($S < 0.6$ to 1 percent) and minimize the use of high-sulfur coal ($S > 3$ percent). If coal-fired boilers in China did not use high-sulfur coal, emissions of SO_2 would be reduced by 1.5 Mt/year by 2010. Although this approach would be environmentally sound, it would require major shifts in social and economic policy. The government would have to close many mines, provide a social safety net for displaced workers, and require power plants to use the more expensive, low-sulfur coal.

Another method of reducing SO_2 emissions is to wash coal before it is burned. This can reduce the sulfur pyrite content of medium- and high-sulfur coal

(S > 1 percent). Washing coal requires minimal investment and entails only a moderate increase in operating costs. However, the associated water pollution must then be addressed. From 1960 to 1980, many engineers and government officials argued that China should use all of its natural resources, including high-ash, low-heating-value coals and mine wastes. Now the country has changed its approach.

If all coals are washed before burning, the desulfurization rate could be as high as 20 percent. This would not only reduce SO_2 emissions, but would also increase the heating value of raw coal and reduce acid deposition; in addition, it would reduce the need for transporting coal. However, it would require approximately 16 billion yuan to increase the amount of coal washed today (280 Mt) to 420 Mt, about 30 percent of the total amount of coal used. With FGD, mine wastes with high sulfur content could also be burned. Waste coal can not be desulfurized at mine sites because it causes acid drainage problems, but it could be desulfurized at the combustion site. Based on the availability and use of waste coal, emissions of SO_2 could be reduced by desulfurization by 6 percent, 1.5 Mt/year. However, power planners have yet to fully analyze the overall costs and benefits, including the social and environmental trade-offs, of coal washing and using low-sulfur coal.

Regulations for SO_2 emissions for the so-called "two control zones of SO_2 emissions in China" were issued in 1998. Under this regulatory regime, coal users are encouraged to use low-sulfur coals, and high-sulfur coal mines must be closed. The regulations also require that all green-field coal-fired power plants with capacities of more than 300 MWe install FGD facilities beginning in 1999. As a result, SO_2 emissions in China have been reduced somewhat (Figure 11). Current regulations also require that all power plants using coal with medium or high sulfur content install FGD facilities by 2010. If wet FGD is used for a plant that produces only 30 GWe, for example, SO_2 emissions will be reduced by 3 Mt. The investment would be at least 33 billion yuan. Investment costs in FGD facilities, at 1,100 yuan per kWe, would be approximately 18 percent of the total investment of power plant capacity.

Another method currently used to control sulfur emissions in China is pulverized coal briquettes with limestone (a desulfurization additive), which is used for most home cooking and heating; the briquettes are manufactured in small neighborhood shops. But briquettes can also be used in larger industrial boilers. With an investment of about 2 billion yuan, 113 Mt of briquettes with desulfurization additives could be produced for domestic and other uses. Even though total operating costs would be high because there are more than 5 million small industrial boilers scattered throughout China, desulfurization could have a major impact on ground-level air pollution, especially in many northern cities where small boilers are ubiquitous.

With the combined use of coal washing, briquetting, circulating fluidized-bed (CFB) boilers, and FGD techniques, total SO_2 emissions could be reduced

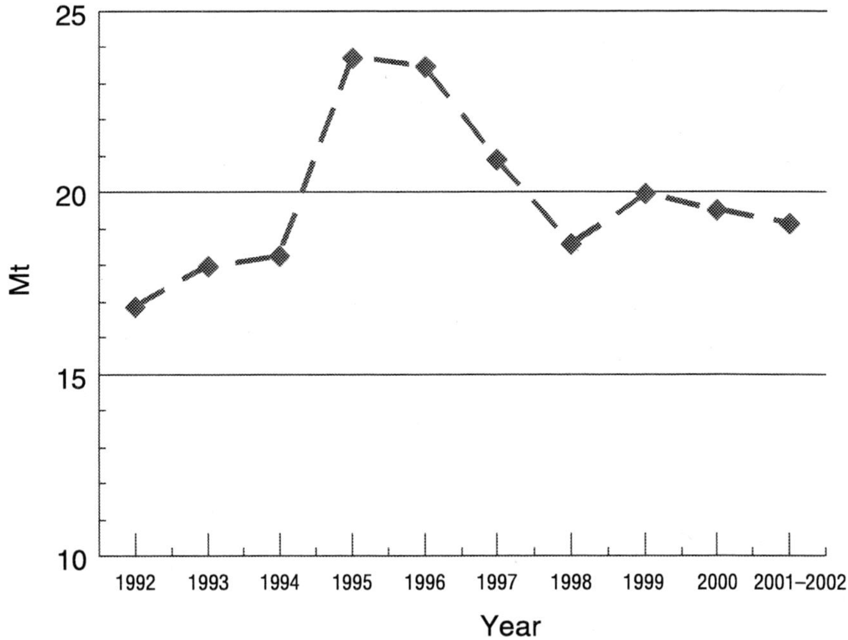

FIGURE 11 Total SO_2 emissions in China.

from the current 30 Mt to 24 Mt. An important element of this program would be to reduce the cost of FGD through massive deployment. Imported wet FGD facilities have sulfur-removal efficiencies as high as 95 percent, but they require a high initial investment (12 to 20 percent of total investment in a power plant) and high operating costs. If the initial investment could be reduced to 8 percent, the same investment would almost double the electricity capacity of coal-fired power plants with FGD. Therefore, research and development on low-cost FGD technology is very important for China.

Although the economic burden would be heavy, if wet FGD were installed in 20 percent of the total coal-fired power capacity in China (400 GWe) by 2010, 80 GWe of capacity would then be produced with FGD. This would reduce SO_2 emissions by 5.6 Mt/year and would require an investment of about 80 billion yuan (based on the 1995 capital cost of 1,200 yuan per kWe). An additional 8 billion yuan would be required for annual operating costs (maintenance, calcium, and electricity consumption). Thus, for the next 10 years, the total costs for wet FGD could be as high as 160 billion yuan. Unfortunately, this sum is considered far too expensive for any developing country.

There have been some positive developments. In the past 10 years, new FGD technologies with Chinese domestic patents have been developed; in addition advanced FGD technologies from abroad have been imported. As a result, the investment costs of wet FGD dropped quickly from 800 yuan/kWe in 2000 to 300 yuan/kWe by the end of 2002. In other words, wet FGD requires only about one-quarter the investment cost required in 1995, a cost many developing countries can afford. The total capacity of coal-fired power plants equipped with wet FGD in China was only 1 GWe at the end of 1995. It is expected to reach 14 GWe by the end of 2006, almost doubling every three years. Capacity could reach 35 GWe by 2010, with a total investment in low-cost FGD systems of 20 billion yuan. Table 4 shows the potential decrease in SO_2 emissions by 2010, with investment and operating costs one-fourth to one-third lower than in 1995.

Although water can be reused and recirculated in wet FGD systems, large quantities of water are lost through evaporation. Even the semidry FGD process, such as the FGD-CFB technique, consumes large amounts of water, generally 70 to 80 percent of the amount consumed for wet-FGD systems. Therefore, these systems may be difficult to use in northern and northwestern China, where water is not readily available. Therefore, China should focus on developing FGD systems that require smaller quantities of water. Tsinghua University and Tokyo University recently initiated a joint project to study dry FGD processes that require less water.

CONTROL OF NITROGEN OXIDE EMISSIONS

The problem of NO_x emissions has become quite serious in China. Emissions are mainly from coal combustion, but also can originate in any high-temperature combustion process, such as internal combustion engines. Although NO_x emissions from coal-fired power plants can be easily controlled, there is no regulatory regime in place. Thus, they totaled 1.3 Mt in 1989, 2.65 Mt in 1995, and 2.85 Mt in 2000. Total NO_x emissions from coal combustion and cars in 1996 was 12 Mt and continue to increase.

NO_x emissions depend largely on energy-consumption levels and GDP, and the highest GDP in China is concentrated mostly in developed areas along the east coast of China. Average annual NO_x emissions in kg/year per capita are generally related to energy consumption per capita, although this relationship does not necessarily hold true in all parts of China. For example, in Shanxi province and Xinjiang, GDP levels are low and energy-consumption levels are low, but NO_x emissions per capita are high. This is because the percentage of energy from coal is much higher in those areas than elsewhere in the country. In those areas, the value of NO_x emissions per 1,000 yuan of GDP is also very high.

Overall, more than 70 percent of NO_x emissions in China are primarily from coal combustion (Figure 12). The highest NO_x level of emissions from coal combustion in China was 9.17 Mt in 1996. Oil combustion and cars accounted for about 15 percent of the overall total. In big cities, such as Beijing, Shanghai, and

TABLE 4 SO_2 Emissions Reductions and Costs by 2010 (Ca/S is the calcium-to-sulfur molar ratio)

				FGD Systems	
	Unit	Coal Preparation	Use of CFB boilers	Cost of Wet FGD before 1995	Low-Cost FGD
Desulfurization efficiency	Percentage	~20	At Ca/S = 2 ~75	At Ca/S = 1.1 ~95	At Ca/S = 1.1 85–95
Required reduction of SO_2 emissions	Mt/year	2	0.4	3.1	3.1
Total investment costs	Billion yuan	~22	~6	~70	~20
	Yuan/kg SO_2	~11	~15	~22.6	~6
Total operating costs (10 years)	Billion yuan	~18–24	~6.4–7.6	~60–70	~22–25
	Yuan/kg SO_2	~0.8–1.2	~1.0–1.6	~1.9–2.3	~0.7–1.1

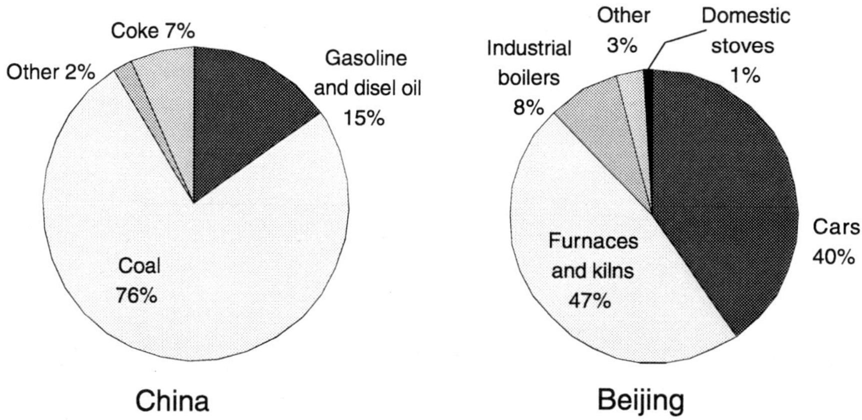

FIGURE 12 NO_x emission sources.

Guangzhou, however, oil combustion and cars accounted for nearly half of the total.

The most widely used technique for reducing NO_x emissions will be low-NO_x coal-combustion techniques, which include CFB boilers and low-NO_x burners. The investment in low-NO_x burners is about 0.03 percent of the capital cost of a power plant and 0.18 percent of the capital cost of a boiler. Over the years, low-NO_x burners have been installed in 130 boilers in 69 coal-fired power plants. In addition, low-NO_x, pulverized-coal combustion techniques have been developed by research institutes and universities in China for different types of coal-fired utility boilers. Chinese utility power plants are now beginning to use modern selective catalytic reduction (SCR) to reduce NO_x emissions.

REDUCING PARTICULATES, DUST, AND CARBON DIOXIDE EMISSIONS

The coal-fired electricity-generating capacity in China has increased rapidly in the past decade. From 1990 to 1999, capacity increased from 76 GW to 180 GW, and annual coal consumption by power plants increased from 240 Mt/year to 400 Mt/year. In the same period, annual dust emissions from power plants decreased slightly (from 3.63 Mt/year to 3.4 Mt/year) and flue-gas dust concentration decreased to 50–150 mg/m^3 as a result of the installation of many four-electric-field ESPs, with dust-collection efficiencies as high as 99.5 percent. These levels are on a par with systems in the United States and European Union.

Despite the decrease in annual dust emissions from power plants, ambient air quality in most Chinese urban areas deteriorated because of an increase in very small, inhalable particles. The diameter of these particles is less than 10μm (PM_{10}), too small to be collected by a locally manufactured ESP. High levels of PM_{10} reduce visibility. Chinese cities that are heavily polluted by PM_{10} included Chongqing, Taiyuan, and Hohhot. Annual PM_{10} values for several international cities are shown for comparison in Table 5.

Inhalable PM_{10} particles in Chinese urban areas are mainly from coal combustion, whereas in Western countries, most urban PM_{10} originates from tailpipe emissions from cars. Only in big Chinese cities, such as Beijing, Shanghai, and Guangzhou, where there are many cars, are emissions of PM_{10} from cars close to emissions from coal combustion. Because the percentage of electricity generation from coal consumption is expected to rise to 65 percent by 2050, controlling PM_{10} from coal-fired power plants is very important, and efforts should be focused on the development of low-cost, high-efficiency dust-collection technologies.

In the past, power plants fueled by residual heavy oil from petroleum refineries were not required to have dust-collection devices. Today, dust concentrations are measured at the exit of every boiler in Beijing, which typically measures 300–400 mg/m^3, generally much higher than after passing through a four-electric-field ESP at a typical coal-fired power plant (50–150 mg/m^3). Another approach to collecting dust is a bag house with a polyporous membrane; dust concentrations with this method average 10 mg/m^3. Thus, PM_{10} pollution from this source has been greatly reduced.

CO_2 is a greenhouse gas, and CO_2 emissions are therefore implicated in global climate change. The most important and enterprising measures for reducing CO_2 emissions are devices with improved energy efficiency and energy saving. Overall, China must keep working on reducing the amount of energy consumed for every kWh of electric power generated. In addition, China must control emissions of heavy metals (e.g., mercury, arsenic, etc.) from power plants.

OPTIMAL ECOLOGY OF COAL ENERGY SYSTEMS

By-products of calcium-sorbent desulfurization systems, $CaSO_4$ and $CaSO_3$, as well as fly ash from coal-fired power plants, can be used as additives in cement or for construction materials for highways. Recently, other uses have been found, such as improving alkali soils, red acid soils, and desert soils.

There are about 100,000 square kilometers of saline-alkali soils in northern and northwestern China. For the past eight years, researchers from the University of Tokyo, using by-products of FGD, have been able to significantly improve alkali soils in the Kangping field of Shenyang. Since 2000, alkali soil in the Tumochuan alkali field close to Hohhot, the capital of Inner Mongolia, has also been significantly improved. After soils have been mixed with FGD by-products, farmers appear to have good harvests of corn (Figure 13), sunflowers, alfalfa, and

TABLE 5 Annual Average PM_{10} Concentrations (December 1997 through March 1998)

	Rochester, UK	Zurich, Switzerland	Fresno, USA	Tokyo, Japan	Milan, Italy	Zingbo, China	Beijing, China	Hohhot, China
PM_{10} (mg/m^3)	0.018	0.031	0.083	0.088	0.099	0.230	0.250	0.541

(a)

(b)

FIGURE 13 Test field of alkali soil in Tumochuan, Inner Mongolia. **a.** Before treatment. **b.** After treatment.

other crops on previously infertile lands. The positive effects appear to last more than eight years without additional treatment. Since 2001, areas of red acid soils in southern China have also been treated. This area is about 900,000 square kilometers, 10 percent of the total territory of China. Test results for growing soy beans, peanuts, radishes, sugar cane, rice, and eucalyptus in treated soils have also been positive.

CONCLUSIONS

Total energy consumption in China is significant, but the average energy consumption per capita is modest compared with consumption in developed countries. Because of its enormous size and rapid growth, Chinese energy policy will have a significant impact on the rest of the world. Most of the energy consumed in China, now and for the next 50 years, will come from coal. Controlling air pollution, therefore, must be a high priority for China. China needs advanced, low-cost pollution-control techniques for industrial plants, power plants, and the thousands of small boilers scattered throughout the country. China also needs low-cost, low-water-consuming FGD techniques, low-cost techniques for reducing NO_x emissions, and efficient, low-cost, control techniques for reducing PM_{10} emissions.

China's development of pollution control and optimal management technologies must go hand in hand with rapid economic development. Therefore, technologies for addressing China's environmental and ecological problems must be suited to the stage of economic development and in keeping with national development goals. China must find ways to optimize coal-energy systems and treating large areas of alkali soils, acid soils, and desert soils. At the same time, China must pursue research and development on energy-conservation measures, renewable-energy systems, and low-emission transportation systems. Both the government and outside actors are assisting in the development of pollution-control systems.

BIBLIOGRAPHY

Chinese Academy of Engineering. 1998. Consultation Report on Energy Stratagem of Sustainable Development of China. Beijing: Chinese Academy of Engineering. In Chinese.

Chinese Academy of Engineering. 2001. The Consultation Report on Efficient Combustion and Pollution Control of Coal-Fired Industrial Boilers in China. Beigjing: Chinese Academy of Engineering. In Chinese.

Development of Clean-Coal Technology

HONGGUANG JIN, RUIXIAN CAI, and BAOQUN WANG
Institute of Engineering Thermophysics
Chinese Academy of Sciences

Clean-coal technology has attracted worldwide attention since the 1980s. China, which ranks first in the world in terms of coal production and consumption, included clean-coal technology in its national sustainable development plan in the 1990s. Clean-coal technology could lead to revolutionary changes in the development of key industries in China, specifically the electric-power industry and the chemical industry.

The Chinese electric-power industry has grown rapidly in recent years. Total installed capacity of electric power is now 350 gigawatts (GW), and this figure is expected to rise to 1,100 GW by 2030. Coal-based power generation will account for about two-thirds of that amount. With current technology, coal consumption is not very efficient and causes serious pollution. Improving the thermal efficiency of power plants by adopting clean-coal technology could go a long way toward addressing both problems.

The most damaging pollutants from coal combustion are sulfur dioxide (SO_2), nitrogen oxides (NO_X), carbon dioxide (CO_2), and particulate matter (PM). China has successfully reduced emissions of SO_2 and NO_X and decreased PM from power plants. But controlling CO_2 emissions and mitigating their greenhouse effect is still one of the most serious environmental challenges facing China, as well as the rest of the world. Up to now, there have been three main approaches to controlling greenhouse gas emissions: (1) improving the energy efficiency of power plants; (2) switching to fuels that produce less CO_2 per unit energy; and (3) using renewable and/or nuclear energy.

Separation and sequestration technology is still not considered by many scholars to be an economically and/or technologically viable solution to the CO_2

problem. A major difficulty in capturing and sequestering CO_2 is the huge volume of the gas stream. Fossil fuels (especially coal) produce a significant amount of carbon per unit energy; thus, a lot of CO_2 must be captured. If the CO_2 is diluted by nitrogen, the volume may be two or three times as large. Recovering this large volume of gas places a huge burden on the system in terms of the high cost and the large amount of energy consumed in the separation process. The significant energy required for CO_2 separation (e.g., steam or refrigeration) greatly reduces the net energy output and efficiency of the power plant. After separation, the CO_2 must be compressed, cooled to a liquid, and transported long distances to a storage area underground or in the ocean. Fully one-third of the energy produced might be consumed in the compression process. Overall, recovery of CO_2 by coal-fired power plants is technologically complex and lowers the efficiency of coal conversion. In the long run, CO_2 capture technology will have to be integrated with thermal physics systems and chemical engineering systems, which will require some technological breakthroughs.

Current clean-coal technologies all offer opportunities for capturing CO_2. These include integrated gasification combined-cycle (IGCC) technology, integrated internal and external coal-fired combined-cycle technology, and coal-based polygeneration systems. These three technologies, as well as a coal-based energy network suitable for China, are discussed below.

INTEGRATED GASIFICATION COMBINED-CYCLE SYSTEM WITH CO_2 RECOVERY

IGCC, a relatively new technology, has drawn worldwide attention because it increases efficiency, reduces pollution, and requires less water than current systems (DOE, 1996). Dozens of IGCC demonstration plants have been established or are being established in the United States and Europe. China is beginning to carry out research on IGCC techniques, including gasification, gas purification, gas turbines, and heat-recovery systems, and is considering establishing its first 300–400 megawatt (MW) IGCC demonstration plant in Yantai.

The inherent advantages of IGCC in reducing pollution are indisputable. Waste emissions (SO_2, NO_X, and PM) from IGCC are lower than from a pressurized, fluidized-bed combustor combined-cycle plant or a pulverized coal-fired power plant with flue-gas desulfurization. The most distinctive characteristic of IGCC is the generation of a concentrated stream of CO_2 with no dilution by nitrogen; pre-decarbonization is also feasible in the IGCC system.

The two critical technologies for IGCC with CO_2 recovery are gasification and water-shift reaction (Figure 1). The synthesis gas (syngas) from the gasifier contains a large portion of CO, which is converted to CO_2 and H_2 through a water-shift reaction:

$$CO + H_2O \rightarrow CO_2 + H_2$$

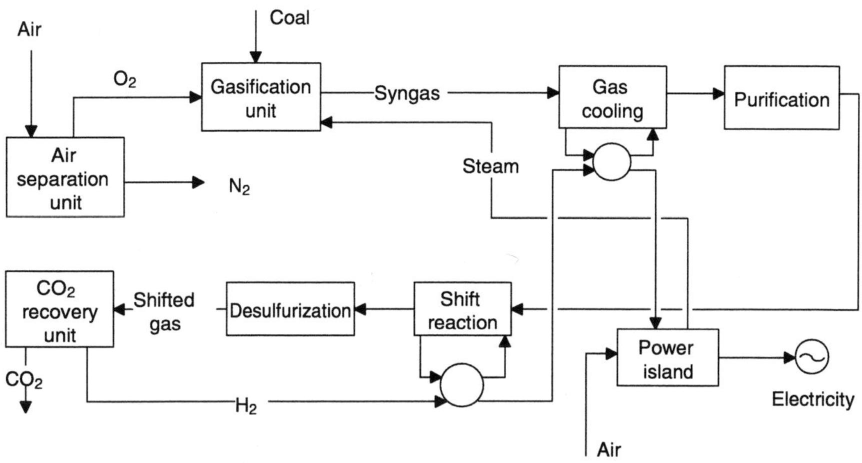

FIGURE 1 Schematic diagram of pre-decarbonization in IGCC. Source: Doctor et al., 1996.

The shifted gas has a high concentration of CO_2. For example, for the air-blown IGCC with Kellogg-Rust-Westinghouse (KRW) gasifier, the CO_2 mole concentration is about 24 percent (in oxygen-blown KRW IGCC systems it can be as high as 40 percent). By comparison, the flue gas from current coal combustion contains only 5 to 6 percent CO_2. Because the volume of the treated gas stream is substantially reduced, the equipment for CO_2 separation costs less, and less energy is consumed (Doctor et al., 1996).

A chemical-absorption method using amine solvents, such as mono ethanol amine (MEA) or methyl diethyl amine (MDEA), is commonly used in the capture of CO_2 from the IGCC system. Other feasible technologies for large-scale CO_2 recovery are hot K_2CO_3 absorption, Selexol, and low-temperature absorption by methanol. Membrane and membrane-reactor separation methods are now being researched. Energy consumption for CO_2 separation by conventional methods in IGCC systems is still rather high, and system efficiency is lowered by about 6 to 10 percent as a result of CO_2 recovery. Table 1 compares the effects of CO_2 separation methods on the efficiency of the IGCC system. In terms of energy consumption, Selexol is the best option for the recovery of CO_2 in IGCC. However, Selexol has some disadvantages, such as expensive solvent and high capital costs for equipment (Doctor et al., 1994).

Current research on reducing energy requirements for CO_2 separation in the IGCC system is concentrated in two areas: (1) the oxygen-blown IGCC system and the separation of CO_2 and H_2; and (2) the selection of a solvent with low energy consumption. In recent years, considerable research has been focused on

TABLE 1 The Effect of CO_2 Separation Methods on the Efficiency of the IGCC System

Systems/ Parameters	Base Case	Selexol	Hot K_2CO_3	MEA	Methanol
Net power output (MW)	454.37	380.78	328.24	310.5	269.22
CO_2 emission rate (kg/kWh)	0.8353	0.1559	0.1805	0.1909	0.2037
Thermal efficiency (HHV %)	39.62	33.42	28.87	27.33	23.76

Source: Doctor et al., 1994.

improving absorption solvents and equipment, and many solvents with low energy consumption and low corrosion rates have been tested and evaluated.

The combination of CO_2 separation and clean-energy generation in the IGCC system would provide clean, efficient coal consumption and would greatly reduce the amount of energy consumed during CO_2 separation. This integrated system would reduce the intermediate processes of separation, increase the efficiency of the power plant by decreasing the amount of energy consumed during separation, and ultimately create synergy between energy generation and the environment. However, integrating the processes will require a breakthrough in system synthesis. We are currently investigating the integration of a cryogenic air-separation unit into an IGCC system for CO_2 separation. The preliminary results show that this would reduce energy requirements for CO_2 recovery by one-third compared to the IGCC system using the conventional absorption method by amine solvent.

INTEGRATION OF INTERNAL AND EXTERNAL COAL-FIRED COMBINED-CYCLE SYSTEMS

The commercialization of IGCC depends largely on two factors: (1) thermal performance; and (2) economics. With the development of the advanced gas-turbine engine, the thermal efficiency of IGCC has reached 42 to 45 percent and is expected to exceed 50 percent in the near future. Therefore, high cost is now the biggest obstacle to commercialization of the system. Carbon conversion in IGCC, which usually exceeds 95 percent, requires large equipment and high-temperature materials. Thus, the cost of the gasifier is still prohibitive.

We have proposed a partial-gasification IGCC that integrates internal and external coal-fired combined-cycle systems in a novel power-generating system (Figure 2). This system contains three main "islands": partial gasification of coal and purification (A–F); internal-fired gas-turbine cycle (H–K); and external-fired

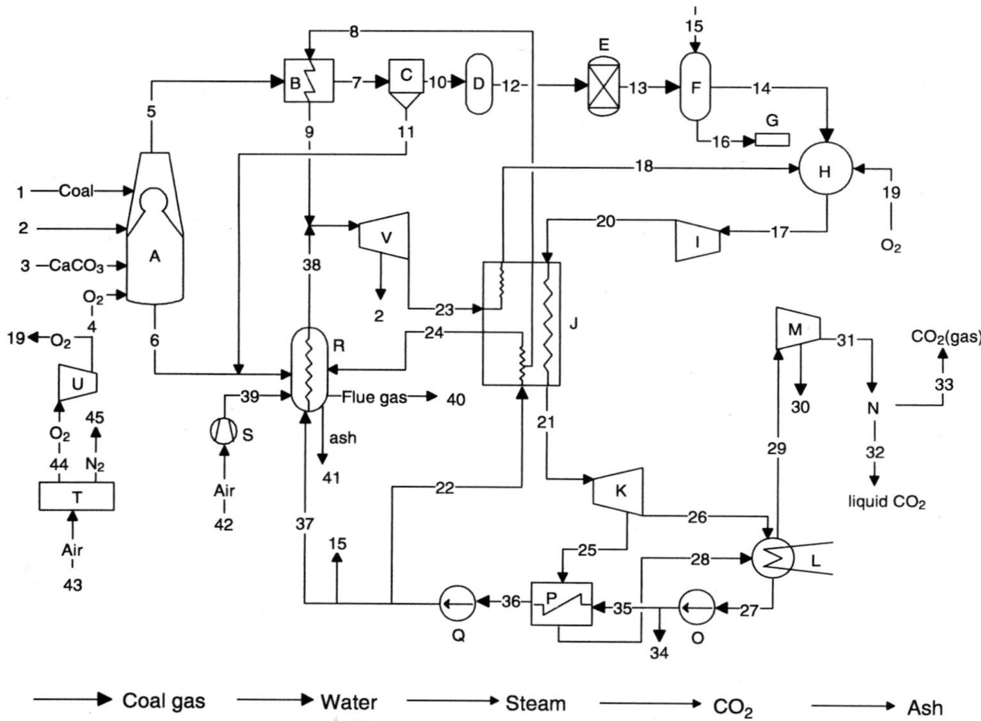

FIGURE 2 Diagram of integration of internal and external coal-fired power-generation combined cycle. A = gasifier. B/P = heat exchanger. C = dust collector. D = hydrolyzer. E = dry desulfurizer. F = washer. G = sewage disposal. H = combustor. I = high-temperature gas turbine. J = reheater. K = low-temperature gas turbine. L = condenser. M = CO_2 compressor. N = refrigerator. O/Q = pump. R = semi-coke boiler. S = air blower. T = air separation unit. U = O_2 compressor. V = high-pressure steam turbine. Source: Gao et al., 2002.

steam-turbine cycle (O–S, V). The associated units are an air-separation unit (T) and a CO_2 recovery unit (M–N). The partial-gasification island provides raw coal gas to the internal-fired gas-turbine cycle and semi-coke to the external-fired steam-turbine cycle, respectively.

In the internal-fired gas-turbine cycle, the raw coal gas is combusted with O_2, and the thermal energy released is used to heat the combustion-gas stream (mainly water vapor), thus maintaining a high cycle efficiency. Because the main components of the combustion product are steam and CO_2, the turbine exhaust can reach vacuum status, taking full advantage of the low temperature of exhaust in the Rankin cycle. CO_2 is separated in the process of condensing the exhaust gas.

In the external-fired steam cycle, high-pressure water is heated to high-pressure steam by the heat recovered from the semi-coke, the reheater, and the high temperature of raw coal gas in the heat exchanger. The high-pressure steam generates energy in the steam turbine (V), and the steam exhaust enters the combustor (H) through the reheater (J).

This novel system, which uses water vapor as the working fluid and a combination of internal and external coal-fired systems, is characterized by high inlet temperature and low discharge pressure. The system has two main advantages over an IGCC system: (1) because partial gasification of coal does not require complete carbon conversion, the gasifier can be substantially smaller; and (2) the temperature for partial gasification is lower than for IGCC; therefore, the equipment is less expensive.

The net efficiency of the novel system is 45.4 percent at an inlet temperature of 1,300°C; the efficiency of IGCC is 45 percent under the same conditions. In addition, the new system recovers about 41 percent of liquid CO_2 simultaneously. By integrating partial coal gasification, advanced thermal cycles, and CO_2 separation, the novel system is more efficient, more environmentally friendly, and has a better economic profile than an IGCC system.

INTEGRATION OF CO_2 SEPARATION AND CLEAN-ENERGY GENERATION BASED ON A POLYGENERATION SYSTEM

Coal gasification is a central technology for both the electric-power industry and the chemical industry. The coal-based polygeneration system, which uses coal gasification and generates syngas, has broad industry applications. The most important feature of polygeneration is the production of methanol, dimethyl ether (DME), and Fischer-Tropsch (FT) liquids, such as FT diesel from syngas.

A polygeneration system can generate various products with high added value, especially clean energy and fuels. In the near term, the products of a polygeneration system could serve a wide range of energy needs with extremely low levels of air pollution. The principle of polygeneration is to integrate and optimize different technologies by compensation and to reduce both the cost of production and pollutant emissions.

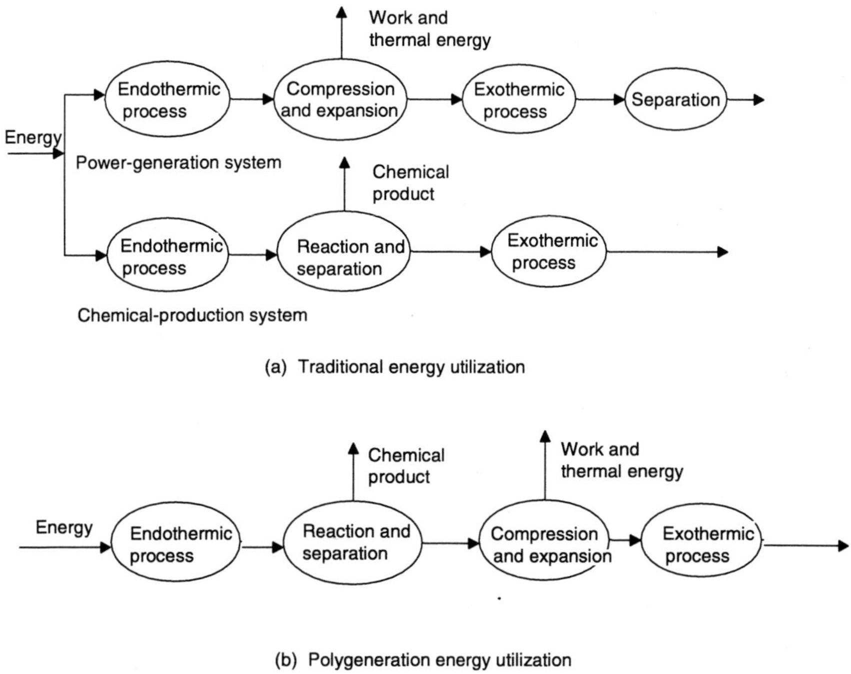

FIGURE 3 Principles of the polygeneration concept. Source: Gao et al., 2002.

The concept of polygeneration is illustrated in Figure 3. In the traditional method of energy consumption (Figure 3a), the power-generation system is independent of the chemical-production system. Because both systems may have processes for the same purpose, such as separation, this is inefficient. In contrast, in a polygeneration system, the power-generation system and the chemical-production system can be integrated (Figure 3b). Because duplications and overlapping processes can be avoided, less energy is consumed. For example, in a traditional power-generating system, contaminations are separated downstream of the system, an energy-intensive process. In the polygeneration system, the reaction and separation are accomplished in one process; thus, chemical production and power generation form a chain. In short, in polygeneration the energy and material exchanges between the power-generation system and the chemical-production system are interdependent. Although polygeneration could be more complex to regulate, the system has great potential for saving both energy and economic resources.

The polygeneration system proposed here integrates CO_2 separation and clean energy (e.g., methanol) generation. The system has once-through, liquid-phase methanol synthesis. As shown in Figure 4, syngas from gasification is

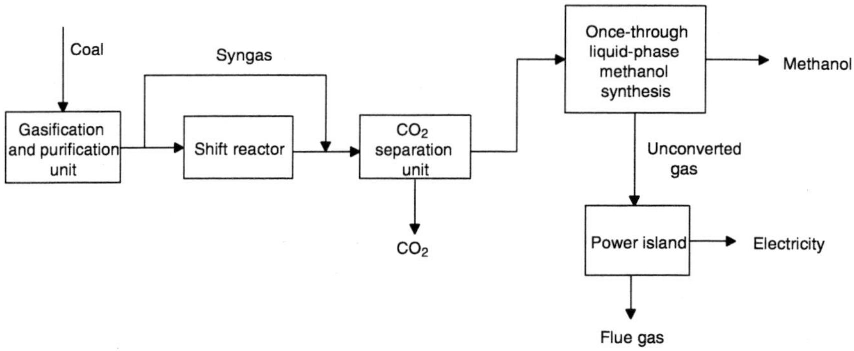

FIGURE 4 Simple block drawing of polygeneration system with integration of CO_2 separation and clean-energy generation. Source: Gao et al., 2002.

divided into two parts: one part that enters the shift-reaction unit and one that serves as regulating gas. Unlike conventional methanol synthesis, which recycles unconverted gas to increase the methanol conversion rate, the polygeneration system controls methanol synthesis by adjusting the flow rate ratio of the shifted gas to the regulation gas and sends the unconverted-gas stream to the power island as fuel. Both clean energy from methanol and electricity are products of the system.

The optimal methanol conversion rate may not be ideal for the overall polygeneration system, however. Thus, acceptable performance will require optimization of both the production of methanol and the production of electricity. This integration will surely affect the CO_2 recovery rate, which is determined by the proportion of syngas sent to the shift-reaction unit.

The integration of CO_2 separation and clean-energy generation could lead to an innovative system that meets both environmental and energy-production requirements. The key component of the polygeneration system is the CO_2 separation unit. The separation of CO_2 from shifted gas changes the proportion of H_2 and CO in the feed gas, leading to a suitable concentration of reactants for methanol synthesis. In other words, the separation of CO_2 and the generation of methanol can be accomplished simultaneously. Thus, integration of CO_2 separation and clean-energy generation eliminates the energy penalty of CO_2 separation.

The integrated polygeneration system has a higher efficiency than the IGCC with CO_2 recovery system, with the same ratio of CO_2 recovery. A comparison of the three systems (Table 2) shows that the thermal efficiency of the integrated polygeneration system is about 5 percent higher than for IGCC without CO_2 recovery and 11 percent higher than IGCC with CO_2 recovery. In addition, the polygeneration system can recover 56 percent of CO_2 at high efficiency. The

TABLE 2 Comparison of the Integrated Polygeneration System, IGCC System, and IGCC System with CO_2 Recovery

Systems/ Parameters	IGCC	IGCC with CO_2 Recovery	Integrated Polygeneration
CO_2 recovery (%)	0	58	56
Thermal efficiency (%)	45	39	50

Source: Gao et al., 2002.

excellent performance of the integrated polygeneration system is attributable to many important features. For example, in the methanol-synthesis process, the cascade use of the chemical energy of syngas increases system efficiency and decreases energy loss from combustion (Gao et al., 2002).

The strategy of the energy industry in China will eventually be to develop a polygeneration system. But for the time being, the power and electricity industry is focusing on the development of a supercritical-steam turbine. Compared with this system, the polygeneration system may require a higher initial investment because polygeneration is a complex system that combines several advanced technologies, including gasification, gas turbines, and high-temperature gas cleanup. However, the efficiency of polygeneration may be as high as 55 to 60 percent if the inlet temperature is 1,500°C. In contrast, the increase in efficiency with the supercritical-steam turbine cycle will be limited to about 45 percent. Moreover, as the advanced technologies mature, there is a good chance that the costs of polygeneration will decrease. In addition, if CO_2 is used for enhanced oil recovery and enhanced coal-bed methane recovery, the higher costs could be offset and would not be an obstacle to the spread of polygeneration.

DEVELOPMENT OF A COAL-BASED ENERGY NETWORK IN CHINA

There is no doubt that the strategic goals of future energy development must be compatible with environmental protection and that clean-coal technology with greenhouse gas control will necessarily be accepted by countries that rely on coal for energy. The obstacle to CO_2 control with current clean-coal technologies is the high energy penalty in the separation of CO_2. Although the capture of CO_2 in the IGCC system can reduce the energy penalty to some extent, so far no such system has been developed. Can we create new-generation clean-coal technologies with CO_2 capture? Will a breakthrough concept be developed that allows coal-dependent countries like China to produce energy in an environmentally benign way? We have attempted to answer these questions based on our research.

For the purpose of answering these questions, we have proposed a coal-based energy network (Figure 5). A polygeneration plant built near a coal mine can convert the coal to power and liquid fuels, such as methanol and DME, which can then be transported to urban areas. These liquids can be used as a raw material for chemical synthesis and clean power generation. The captured CO_2 can be sequestered or potentially recycled into useful products.

The energy network has several distinctive characteristics. First, it is an environment-friendly system that is highly integrated with emerging technologies, such as polygeneration and clean synthetic-fueled multiutilization systems. By taking advantage of these advanced technologies, the energy network can achieve high efficiency and save resources. Second, the system should contribute to CO_2 capture technology because the energy network can capture CO_2 without the energy penalty associated with current systems. This means that the efficiency of the energy network with CO_2 removal may be even higher than the efficiency of the IGCC system, with or without CO_2 capture. Third, the transportation of clean energy requires less energy than the transportation of coal. Finally, the energy network is flexible; it has multi-output upstream and multi-utilization of products downstream. Upstream, the polygeneration network

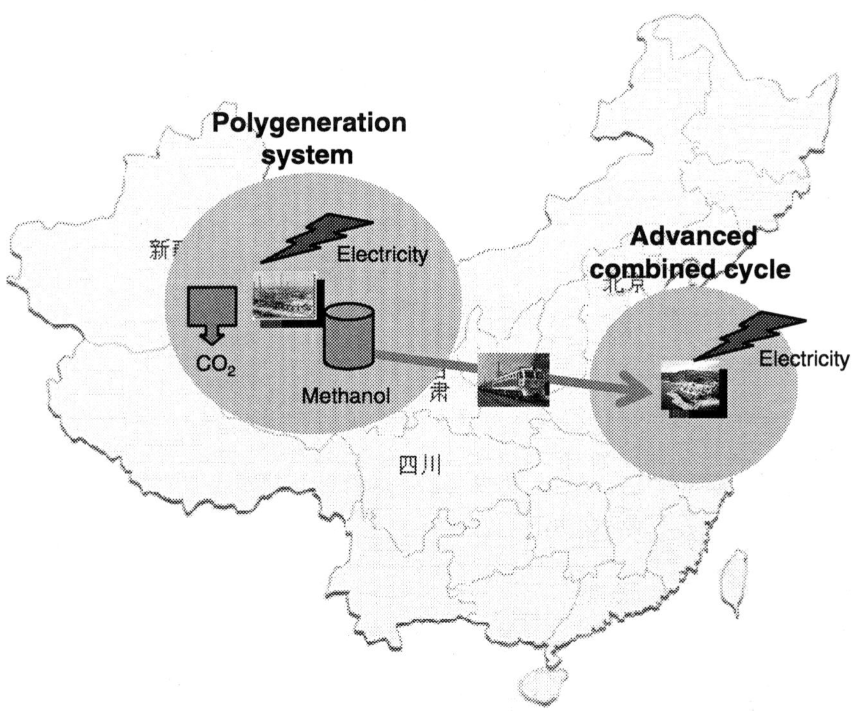

FIGURE 5 Block diagram of coal-based energy network. Source: Gao et al., 2002.

generates both electricity and clean synthetic fuels; the upstream output of IGCC is just electricity, although it has high thermal efficiency. Downstream, clean synthetic fuels have many uses:

- **Power generation.** Clean synthetic fuels can be used as fuels in combined cycles. Most clean synthetic fuels are organic; the combustion products are CO_2 and H_2O, and the CO_2 can be easily recovered by condensing the water vapor. Some clean synthetic fuels (e.g., methanol) effectively use low-temperature thermal energy.
- **Fuel-cell vehicles.** Clean synthetic fuel of methanol has the potential to be used in fuel-cell vehicles. In the future, the widespread use of fuel-cell vehicles will be an important contributor to CO_2 control.
- **Substitute for oil.** Methanol, for example, is a high-octane fuel that can replace gasoline in spark-ignition internal-combustion engines. DME can be used as a substitute for diesel oil in compression-ignition engines to improve energy security in China.
- **Material for chemical synthesis.** The cost of generating clean synthetic fuels from polygeneration might be much lower than from the traditional chemical system; thus, the cost of products derived from clean synthetic fuels may be competitive.
- **Energy for buildings.** Clean synthetic fuels used for central heating systems will require 20 to 30 percent less energy because of higher boiler efficiency.

A CLEAN SYNTHETIC-FUELED POWER-GENERATION SYSTEM

The clean synthetic-fueled power-generation system is a type of combined-cycle system that would use clean synthetic fuel indirectly. The most significant benefit of such a system is that the clean synthetic fuel could be decomposed into synthesis gas (containing CO and H_2) by upgrading the low-temperature thermal energy to chemical energy. Therefore, when the synthesis gas is combusted, the energy output would be higher than from the direct combustion of clean synthetic fuel. A clean synthetic-fueled power-generation system would be much more thermally efficient than a conventional combined-cycle system. At a turbine inlet temperature of 1,300°C, for example, the thermal efficiency of a methanol-fueled power-generation system with intercooling is 60 percent. This is significantly more efficient than the IGCC system without recovery of CO_2, which averages 42 to 45 percent, and even more efficient than the present natural-gas combined-cycle system, which averages 55 percent (Jin et al., 2003).

This estimate is based on the subsystems described above. Upstream, the thermal efficiency of a coal-based polygeneration system, with half of CO_2 recovery ratio, is 50 percent. Downstream, thermal efficiency could be as high as 60 percent with the methanol-fueled power-generation system. For simplification, we have assumed the same energy requirements for transporting coal and methanol. Based on these values, the thermal efficiency of the energy network would be 44 percent, with CO_2 recovery of 60 percent and a CO_2 emission rate of about 0.29 kg-CO_2/kWh.

CONCLUSION

An integrated energy network could address the energy efficiency and environmental problems caused by the use of coal in China. In the long term, this energy network would be superior to current advanced power technologies, such as a supercritical power plant. For a supercritical power plant with 60 percent CO_2 recovery and 10 percent parasitic losses, the average thermal efficiency would be 13 percent lower than for the energy network with the same coal input. Thus, the integrated energy network would be about 40 percent more efficient. In terms of controlling greenhouse gas emissions, the CO_2 emission rate for the supercritical power plant with CO_2 recovery of 60 percent is 0.46 kg-CO_2/kWh. The CO_2 emission rate of the energy network (0.29 kg-CO_2/kWh) would be 37 percent lower with the same ratio of CO_2 recovery because of the increase in thermal efficiency.

Energy consumption and CO_2 emission rates in the energy network could potentially be reduced even further. A rough estimate shows that a 50-percent energy saving would be possible with a higher performance gas turbine and lower energy consumption for methanol synthesis and CO_2 recovery. Studies indicate that the price of methanol and electricity would be lower than when methanol is synthesized from natural gas by individual processes. In addition, producing methanol from coal will probably increase the use of natural gas, which currently accounts for less than 3 percent of energy consumption in China.

In conclusion, an integrated energy network would simultaneously save energy and control greenhouse gas emissions in China.

REFERENCES

DOE (U.S. Department of Energy). 1996. The United States of America and the People's Republic of China Experts Report on Integrated Gasification Combined-Cycle Technology (IGCC). DOE/FE-0357. Available online at: *http://www.lanl.gov/cgi-bin/byteserver.pl/projects/cctc/resources/pdfsprog/clncoalgnrl/M98001976.pdf* (accessed May 14, 2004).

Doctor, R.D., J.C. Molburg, P.R. Thimmapuram, G.F. Berry, and C.D. Livengood. 1994. Gasification Combined Cycle: Carbon Dioxide Recovery, Transport, and Disposal. ANL/ESD-24. Argonne, Ill.: Argonne National Laboratory.

Doctor, R.D., J.C. Molburg, and P.R. Thimmapuram. 1996. KRW Oxygen-Blown Gasification Combined Cycle: Carbon Dioxide Recovery, Transport, and Disposal. ANL/ESD-34. Argonne, Ill.: Argonne National Laboratory.

Gao, L., H. Jin, and Z. Liu. 2002. Energy Analysis of Coal-Based Polygeneration System for Power and Chemical Production. Pp. 137–144 in Proceedings of the 15th International Conference on Efficiency, Costs, Optimization, Simulation (ECOS) 2002. Berlin: Technische Universität Berlin, Institute for Energy Engineering.

Jin, H., H. Hong, and R. Cai. 2003. A Chemically Intercooled Gas Turbine Cycle for Recovery of Low-Temperature Thermal Energy. Pp. 1025–1032 in Proceedings of the 16th International Conference on Efficiency, Costs, Optimization, Simulation (ECOS) 2003. Copenhagen: Ikon Document Services.

Institutional Issues

Environmental Institutions in China

HUA WANG
Development Research Group
World Bank

CHANGHUA WU
Green Development Institute

Air pollution in China is a very serious problem. While national pollution survey data show that total emissions of major air pollutants, such as sulfur dioxide (SO_2), soot, and industrial fugitive dust, peaked in the mid-1990s and have been falling off ever since, concentrations of fine particulates and SO_2 in many Chinese cities remain among the highest in the world. Although metropolitan areas, with a total population of 100 to 200 million, have the most serious air pollution,[1] hundreds of millions of people in rural areas are affected by indoor air pollution from solid fuels used for cooking and heating. Acid rain and acid deposition have also had detrimental effects on regions all over the country. In 2002, acid rain was recorded in more than 90 percent of the cities in acid rain control areas.

Energy, heavy industry, and heating and cooking have been the traditional sources of air pollution in China. The large-scale, inefficient consumption of poor quality and poorly prepared coal in the industrial and energy sectors led to large emissions of pollutants. Pollution from small-scale heating and cooking used to fill the streets of Chinese cities and towns with thick, sticky smoke saturated with soot and a mixture of other pollutants. Recently, however, there has been a distinct shift in the sources of air pollution.

The number of vehicles in China has been growing at a rate of 20 percent per year in many urban areas, and vehicular emissions are fast becoming the primary

[1]In 2002, barely one-third of China's 343 monitored cities were in compliance with the nation's residential ambient air quality criteria.

source of air pollution in cities.[2] In addition, construction, which is widespread and poorly regulated in China, is the source of huge emissions of total suspended particulates (TSP), and in some localities, more dangerous pollutants—although there are few data to describe this phenomenon. Ecological destruction is another main contributor to high levels of TSP in much of northern China, where sandstorms have become more frequent in Beijing and other cities.

China has done many things right in the environmental arena over the past decade. However, much more remains to be done to address the growing environmental challenges. This paper reviews the institutional aspects of environmental protection in China and offers recommendations for improving those institutions. Major recommendations include: (1) putting the quality of life, including environmental quality, at the top of the political agenda in China; (2) developing a national environmental network, with the State Environmental Protection Administration (SEPA) as its anchor, and expanding SEPA's environmental coordination function; and (3) enacting laws requiring the disclosure of environmental information and ensuring public participation in environmental management and the promotion of environmental nongovernmental organizations (NGOs).

The next section of this paper describes the responsibilities of major actors in environmental protection in China. This is followed by sections on regulation, institutional progress, and the future of pollution control in China.

MAJOR ACTORS AND RESPONSIBILITIES

The Legal Foundation

Constitutionally, the National People's Congress (NPC) and its Standing Committee exercise the leading power of the state in China. All administrative, judicial, and procuratorial organs of the state are created by the NPC, to which they are responsible and under whose supervision they operate. China has four major categories of governing institutions—the NPC, the State Council, the People's Court (judicial), and the People's Procuratorates (legal).

According to the Constitution, the state is responsible for protecting and improving the living conditions of the people and the ecological environment. The state also prevents and controls pollution and other public hazards (Article 26), ensures the rational use of natural resources, and protects rare animals and plants. Appropriating or damaging natural resources by any organization or individual by whatever means is prohibited (Article 9).

The Law of Environmental Protection further stipulates that "the plans for

[2]Estimates show that by 2010 in Shanghai, 75 percent of total oxides of nitrogen (NO_x) emissions, 94 percent of total carbon monoxide (CO) emissions, and 98 percent of total hydrocarbon emissions will be from vehicles.

environmental protection formulated by the state must be incorporated into national economic and social development plans; the state [must] adopt economic and technological policies and measures favorable for environmental protection so as to coordinate the work of environmental protection with economic construction and social development" (Article 4).

Article 5 states that "the state shall encourage the development of education in the science of environmental protection, strengthen the study and development of the science and technology of environmental protection, raise the scientific and technological level of environmental protection and popularize scientific knowledge of environmental protection."

According to the law, the competent department of environmental protection administration (i.e., SEPA) under the State Council shall conduct unified supervision and management of environmental protection throughout the country. The competent departments of environmental protection administration (environmental protection bureaus, or EPBs) of local people's governments at or above the county level are responsible for unified supervision and management of environmental protection activities in areas under their jurisdiction.

The state administrative department of marine affairs, the harbor superintendency administration, the fisheries administration and fishing harbor superintendency agencies, the environmental protection department of the armed forces, and the administrative departments of public security, transportation, railways, and civil aviation at various levels shall, in accordance with the provisions of relevant laws, supervise and manage the prevention and control of environmental pollution. The competent administrative departments of land, minerals, forestry, agriculture, and water conservancy of the people's governments at or above county level shall, in accordance with the provisions of relevant laws, "conduct supervision and management of the protection of natural resources" (Article 7).

The 1996 Decision of the State Council on Several Issues Concerning Environmental Protection affirms the importance of public participation and the development of NGOs. The decision states that "a mechanism for public involvement shall be established. Social organizations shall play their role. The public shall be encouraged to [become] involve[d] in environmental protection and to charge against or disclose any kind of illegal activities of violating environmental laws and regulations."

Major Players

Government continues to play the leading role in protecting the environment in China, although participation by an emerging civil society is increasing, and decision makers are working more with the private sector and organizations outside the government.

TABLE 1 Government Agencies with Environmental Responsibilities under State Council

Responsibilities	Actors	Examples
Macro-coordination and control	State Development and Reform Commission	• state land use and protection • water resources and environment planning • nation-wide ecological/ environmental construction plan • coordinate all economic sectors' development
	Ministry of Finance	• fiscal policy, control of state-owned capitals
Specialized public agency	State Environmental Protection Administration (SEPA)	• pollution control • ecosystem protection • natural resource management • urban environmental management
Pollution control	SEPA	
	Ministry of Construction	• solid waste management, wastewater treatment
	Ministry of Railways	• railway industrial pollution control
	Ministry of Communications	• marine (shipping) pollution control
	Ministry of Water Resources	• water quality, water body sink capacity • total water pollutant control
	Ministry of Health	• drinking water quality

Government Players

Government has three major responsibilities: legislative, administrative, and judiciary. The NPC Environment and Resources Protection Committee drafts new environmental legislation and revises existing legislation; local people's congresses are responsible for local legislation. The Supreme People's Court and local people's courts at different levels exercise judicial powers in accordance with law, independently and not subject to interference by administrative organs, public organizations, or individuals. The Supreme People's Procuratorate and local people's procuratorates exercise power independently.

The responsibilities of the State Council (or the Central Government) are distributed among different ministries and agencies (Table 1). Areas of

TABLE 1 Continued

Responsibilities	Actors	Examples
Ecosystem protection	SEPA	
	Ministry of Agriculture	• grassland, wetlands, agricultural biodiversity • aquatic wildlife protection
	State Forestry Administration	• terrestrial biodiversity, land use, forest eco-construction • forest resource protection, land greening
	Ministry of Land and Resources	• land, mineral and marine resources
Natural resource management	SEPA	
	Ministry of Land and Resources	• mineral and marine resource management
	Ministry of Water Resources	• water supply
	Ministry of Agriculture	• fishery, renewable energy, land use
	State Forestry Administration	• timber and forest products • terrestrial wildlife resource use, forest farms, plantations
Others	Ministry of Science and Technology	• R&D
	Ministry of Education	• environmental education
	General Auditing Agency	• environmental auditing
	General Customs Agency	
	Taxation	

responsibility include macro adjustment, coordination, and control (e.g., the State Development and Reform Commission, SDRC) and specialized management areas (e.g., Ministry of Construction, Ministry of Water Resources, Ministry of Land and Resources, etc.) Some agencies are directly affiliated with the State Council, such as SEPA and the State Forestry Administration.

Nongovernmental Organizations

Although China's legal framework does not promote the creation of NGOs in the Western sense, environmentally oriented NGOs are beginning to provide a vehicle for public expressions of environmental concerns and for effecting change, although on a limited scale. Many Chinese NGOs have official governmental sponsors to guide them through the registration process and provide support for their work, political and otherwise. There are two kinds of environmental NGOs in China: (1) fairly independent NGOs and (2) government-organized NGOs.

The activities of many fairly independent NGOs are directed toward environmental education and community development. Friends of Nature, for example, focuses primarily on raising environmental awareness among the public; the group also promotes wildlife and habitat conservation. Green Earth Volunteers is an example of a volunteer organization. Global Village, although registered as a company, focuses on two main areas: (1) production of a series of TV programs; and (2) promotion of residential reuse and recycling in Beijing. In addition, many grassroots and community-level NGOs scattered around the country are working on specific issues in their communities.

A number of international NGOs have also established their presence in China. They include the World Wildlife Foundation-China, Ford Foundation, Oxfam, Leadership for Environment and Development (LEAD) International, and the Nature Conservancy.

Government-organized NGOs are usually state sponsored and are often established by state agencies or well known Chinese leaders or retired officials. Often large, national-level organizations that receive a large part of their funding from the government (although increasingly from other sources as well), government-organized NGOs are not grassroots organizations. They are focused on an elite audience of scholars, policy makers, and government officials. The clear advantage of these organizations is their ability to bring together scholars and officials from a wide range of institutions that normally find it difficult to interact in China's highly vertical bureaucratic structure.

Institutes and Think Tanks

The explosive increase in international academic exchanges and the involvement of Chinese experts in cooperative environmental development projects have created an atmosphere of more independent, critical thinking among research institutes and think tanks. For instance, the Rural Development Institute of the People's University of China in Beijing is one of many educational institutions that combine academic research and rural development projects in China's interior. The Beijing Environment and Development Institute (BEDI) conducts applied research on environmental issues and encourages market-based solutions to environmental problems. Regionally based institutions, such as the Chinese Academy of Sciences Institute of Geography and the Northern Normal University Institute of Environmental Sciences, have developed impressive research and training programs to address specific regional needs.

The Beijing-based South-North Institute for Sustainable Development has conducted some very successful small demonstration projects on biogas energy for small farmers in the Baima Snow Mountain nature reserve in Sichuan Province. The prestigious Energy Research Institute, which is under the SDRC umbrella and is supported by the dynamic Center for Renewable Energy Development, is an example of a think tank that has had a good deal of influence over China's energy policies.

A number of institutions have also been actively involved in the development and expansion of environmental management systems in industry. The China National Cleaner Production Center, under SEPA, is an example of an institution that advises businesses on environmental auditing, cleaner production, and setting up ISO 14000 systems.

International Actors

Global changes in ecology, economics, and politics have led to a shift of some power and authority away from nation states and toward supranational, regional, and local levels of governance. International environmental conventions and agreements and multilateral, bilateral, and regional cooperation are a few examples. China has been actively involved in the preparation and implementation of international environmental conventions and now has obligations under more than 80 bilateral and multilateral environmental treaties (see Table 2). Former Premier Zhu Rongji reinforced China's commitment to reducing greenhouse gases at the World Summit on Sustainable Development in September 2003, when he announced that China was moving forward with ratification of the Kyoto Protocol of the Framework Convention on Climate Change.

China is committed to continue taking steps to protect environmental resources. These steps have included: current state assessments; sharing of information; participation in international cooperative activities; preparation of national action plans; and integration of environmental protection and general development plans and programs.

Because China is held accountable to the international bodies (such as the United Nations Environment Program, UNEP) in charge of implementing these agreements, the ratification and implementation of global conventions has provided a strong incentive for placing environmental issues high on the national political and development agenda. In addition, these accords often provide financial and other support for the development and implementation of environmental laws by domestic institutions.

China is also involved in multilateral, bilateral, and regional cooperative projects, as well as projects with NGOs. The World Bank and the Asian Development Bank have been important in promoting China's environmental protection and sustainable development activities.

TABLE 2 Major International Environmental Conventions Ratified by China

Convention	Date of Ratification	Leading Government Agencies
UN Framework Convention on Climate Change	2002	Ministry of Foreign Affairs, SDRC, MOST, Agriculture, Finance, SEPA, Forestry, Land/Resources, etc.
Stockholm Convention on Persistent Organic Pollutants	2001	SEPA, Foreign Affairs, Agriculture, SDRC, Health
UN Convention to Combat Desertification	1997	Foreign Affairs, Forestry
UN Convention on Biological Diversity	1992	SEPA, Foreign Affairs, Finance, Agriculture, Forestry
Montreal Protocol on Ozone Depleting Substances	1991	SEPA, Foreign Affairs, Agriculture, Finance
Convention on Preventing Ocean Pollution by Dumping Wastes and Other Materials	1985	State Oceanic Agency, Communications, Foreign Affairs
Convention on Internationally Important Wetlands Especially As Water Fowl Habitats		Foreign Affairs, Water Resources, SEPA, Forestry
Framework Convention on Tobacco Control		SDRC, Health, Foreign Affairs, Finance
Convention on International Trade of Endangered Wildlife Species	1981	Forestry, Foreign Affairs, Agriculture
Protocol on Environmental Protection to the Antarctic Treaty	1991	
Basel Convention on Transportation of Hazardous Wastes	1995	

Administrative Oversight

After the major administrative reforms of 1998, the State Council issued the Three-Determinations Program—emphasizing functions, staff, and personnel. Through this program, each government agency has been required to establish clearly defined functions, responsibilities, and powers and to specify the number of staff members to be responsible for each function.

State Environmental Protection Administration

SEPA is authorized to conduct "unified supervision and management of environmental protection throughout the country." Under the Three-Determinations Program, SEPA has responsibilities in the following areas:

- **Policy and regulatory functions:** drafting guidelines, policies, laws, regulations and administrative regulations; providing environmental assessments for economic and technological policies; economic and development planning, with the endorsement of the State Council; drafting a national plan for environmental protection; and drafting and supervising implementation of plans in specific areas such as pollution prevention and control and ecological protection for key regions and basins identified by the state.
- **Enforcement and supervision:** organizing the enforcement of various pollution-prevention and control laws and regulations; guiding, coordinating, and supervising marine environmental protection activities; overseeing the use of natural resources that effect the ecological environment; overseeing the environmental protection of natural reservation zones, scenic and historic sights, and forestry parks; supervising efforts to protect biodiversity and manage natural resources (e.g., prevention and control of desertification); making suggestions for the establishment of new national natural reservation areas; and supervising national natural reservation areas.
- **Cross-cutting and regional issue coordination:** directing efforts to deal with interdepartmental and interprovincial issues; investigating and dealing with major pollution accidents and ecological damage; resolving interprovincial disputes; organizing and coordinating efforts to prevent and control water pollution in national key basins; taking charge of the environmental supervision and administrative examination; and organizing the examination on the enforcement of environmental protection laws.
- **Environmental standards:** formulating national standards for environmental quality and discharges; documenting local standards; reviewing comprehensive urban plans for environmental protection; and publishing annual environmental protection reports and related documents (e.g., a national sustainable development profile).
- **Environmental management and environmental impact assessment (EIA):** formulating and organizing the implementation of regulations for management of environmental protection; examining and approving the EIA of major new construction; directing the comprehensive control of urban and rural environmental protection; and directing efforts for rural environmental protection and the construction of national ecological pilot areas; and promoting eco-agriculture.
- **Research and development (R&D), certification, and environmental industry:** coordinating environmental science and technology,

important research projects, and technological pilot engineering; overseeing the development of the environmental protection industry; managing the national certification of environmental management system and environmental labels; and developing and organizing the qualification approval system.

- **Environmental monitoring and information disclosure:** supervising monitoring, statistics, and information gathering; formulating the regulations and norms for monitoring; organizing the establishment and management of a national monitoring network and national information network; organizing, overseeing, and coordinating environmental education and publicity; and promoting the participation of the public and NGOs in environmental protection.
- **Global environmental issues and international conventions:** formulating national principles for addressing global environmental issues; managing international cooperative efforts; helping to coordinate major events on environmental protection; participating in the negotiations of international environmental protection agreements; coordinating the implementation of international environmental agreements; coordinating and implementing foreign-funded environmental projects; dealing with foreign affairs in relation to environmental protection with the endorsement of the State Council; and communicating with international environmental organizations.
- **Nuclear safety:** managing radioactive wastes and nuclear materials; formulating guidelines, policies, laws, and standards; participating in emergency responses to nuclear accidents and radioactive accidents; supervising and managing the safety of nuclear facilities; developing mineral resources with radioactive content; and supervising the safety of nuclear pipelines and pressure-bearing facilities.

Since SEPA was elevated to the status of a ministry, it has been under the direct leadership of the State Council, from which it receives almost all of its funding. SEPA's national policy activities are supported by the Committee on Natural Resources and Environmental Protection, under the chief legislative body—the NPC.

State Development and Reform Commission

The SDRC, originally the State Development and Plan Commission, is the product of the latest government restructuring symbolizing China's final move from a command-and-control economy to a market economy. SDRC is now the central agency responsible for macro-coordination and control of the market economy.

In the areas related to environmental protection, the SDRC is responsible for coordinating the policies of land use, exploitation, and protection; participating in the development of plans for balanced use of water resources and environmental protection; leading the country toward sustainable development; and directing the work of the National Climate Change Countermeasure Coordination Group. The SDRC is also in charge of linking and balancing the development of agricultural, forestry, water resource, meteorological, aquatic products, husbandry, and land cultivation policies. As part of the 2003 reform, the SDRC took over the responsibilities of the dissolved State Economic and Trade Commission for the development of policies and planning for economic sectors, including energy, transportation, and raw materials.

Other Ministries and Agencies with Environmental Responsibilities

Several other government agencies have environmental responsibilities, many of which were summarized in Table 1. The Ministry of Foreign Affairs is involved in the negotiations and ratification of all international conventions and other agreements. The Ministry of Water Resources (MOWR) has responsibilities related to river basin management and water management outside of urban areas. The Ministry of Construction has a strong influence on and responsibilities in urban environmental infrastructure development. The Ministry of Agriculture (MOA) has overall responsibility for most of what happens outside of urban areas, including the regulation of township and village enterprises. The Ministry of Science and Technology (MOST) is deeply involved in sustainable development and climate change negotiations and has responsibilities in technology R&D. The State Forestry Administration (SFA) has significant responsibilities related to forest and wetlands management and to the preservation of biodiversity. The Ministry of Land and Resources is responsible for the planning, administration, protection, and rational use of land, mineral, and marine resources, including mapping and cadastral management. The State Ocean Administration (in the Ministry of Land and Resources) is responsible for management of coastal and marine waters, including biodiversity conservation. The State Meteorological Administration takes part in climate change negotiations, and has responsibilities in regional air quality management. The Ministry of Communications shares responsibilities with SEPA for vehicle emissions control, the implementation of which is the responsibility of public security bureaus. The General Auditing Agency is responsible for environmental auditing and related activities.

Local Agencies

Almost all national agencies have counterparts at the provincial, municipal, and county levels. For example, SEPA's counterparts are EPBs or offices (EPOs) down through the administrative hierarchy at the provincial, municipal, county,

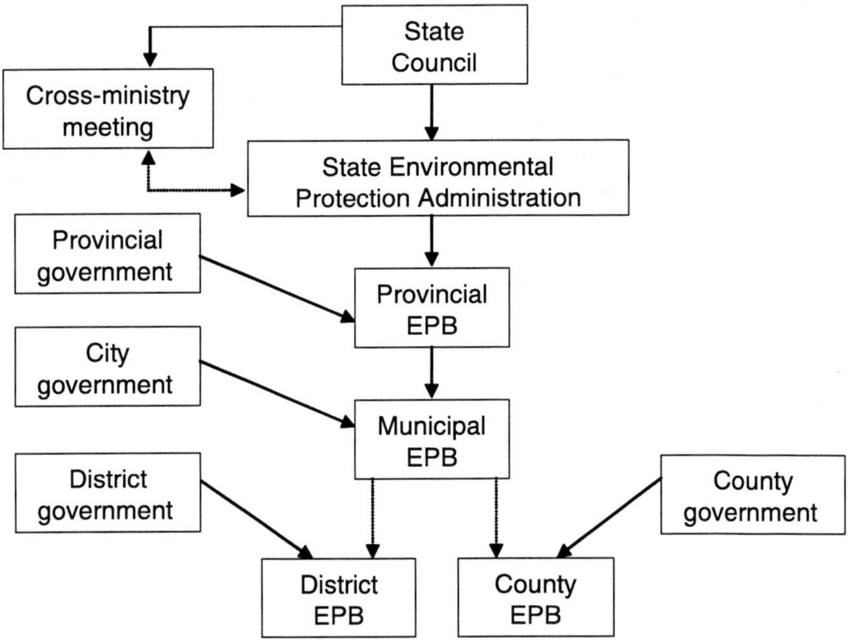

FIGURE 1 Environmental protection apparatus. The Environmental Protection Bureaus (EPBs) administer, supervise, and manage environmental protection activities at the various local levels.

district, and in some places, township levels (see Figure 1). The chief responsibility of these local environmental units is to enforce laws and implement policies designed by SEPA and to draft local regulations to supplement central regulations. EPBs work directly with local factories and other polluters, as well as with industry bureaus and local governments.

REGULATION

Policy Making

According to the Law on the Prevention and Control of Air Pollution, the national government shall adopt measures to control or gradually eliminate all emissions of air pollutants throughout the country in a planned way; the local governments at all levels shall be responsible for ambient air quality in the areas under their jurisdiction and shall formulate plans and adopt measures to ensure that ambient air quality in their jurisdictions meets defined standards.

SEPA and local EPBs shall exercise unified supervision and management of the prevention and control of air pollution. The public security, communications, railway, and fishery departments at all levels shall, according to their respective functions, be responsible for the supervision and management of air pollution by motor vehicles and vessels.

SEPA is responsible for setting national ambient air quality standards. Governments at the provincial level may establish their own local, supplementary standards for items not specified in the national ambient air quality standards and report to SEPA for documentation.

In accordance with the national ambient air quality standards and the nation's economic and technological conditions, SEPA shall establish national air pollutant emission standards. The governments at the provincial level may set local standards for items not specified in the national air pollutant emission standards, as long as they are more stringent than the national standards. Local standards should be reported to SEPA for the record. All emissions of air pollutants are subject to local standards where they have been established.

Indoor air pollution has traditionally been dealt with by public health and agriculture agencies and is not on SEPA's agenda. Because public health agencies are being overwhelmed by other public health issues and agricultural agencies are facing redefined missions, there is a danger that indoor air quality may be left out of the government agenda completely. As China's top environmental agency, SEPA should lead efforts to coordinate activities to address indoor air pollution.

According to the Law on Prevention and Control of Water Pollution, competent departments (SEPA) under the State Council and local governments at various levels shall incorporate the protection of the water environment into their plans and adopt ways and measures to prevent and control water pollution.

SEPA and environmental protection departments at all levels shall exercise unified supervision and management of the prevention and control of water pollution. Navigation administrative offices of transportation departments at various levels shall exercise supervision and management of pollution from vessels. Water conservancy departments, public health departments, geological and mining departments, municipal departments, and water sources protection agencies for major rivers at various levels shall supervise and manage the prevention and control of water pollution by performing their respective functions in conjunction with environmental protection departments.

SEPA shall establish national ambient water quality standards. The governments of provinces, autonomous regions, and municipalities may establish local, supplementary standards for items not specified in the national ambient water quality standards and report the same to SEPA for the record.

In accordance with the national ambient water quality standards and the country's economic and technological conditions, SEPA shall establish national pollutant discharge standards. The governments of provinces, autonomous regions, and municipalities may establish local standards for the discharge of water pollutants not specified in the national standards; for items specified in the national standards, local standards must be more stringent than national standards. The local standards for the discharge of water pollutants must be reported to SEPA for the record.

Policy Instruments

Policy instruments in China include command-and control, market-based, information-based, and voluntary approaches (see Table 3). China has continued to improve its capacity in air pollution prevention and control. The first air pollution control law went into effect in 1987 and was amended in 1995 and 2000 to strengthen enforcement and address emerging issues. The 2000 amendment includes several new mandates that will improve future air quality management, such as the endorsement of emission fees and emission permits, both of which are potentially important regulatory instruments. China has also made significant adjustments in its fuel mix through large investments in both residential fuel switching and energy efficiency through technical innovation and structural adjustment. These pro-environment energy policies are significant additions to the air quality improvement agenda. The government has already developed a national strategy for reducing motor vehicle emissions based on international experiences and China's situation. Key components include: phasing out leaded gasoline; tightening emission standards for all categories of new vehicles; upgrading vehicle inspection and maintenance programs; adopting cleaner fuels; and implementing traffic and demand management.

TABLE 3 Pollution Control Instruments

Command-and-Control Instruments	Economic Incentives	Voluntary Instruments	Public Disclosure Instruments
• Concentration-based pollution discharge limits • Mass-based controls on total provincial discharge (pilot only) • Environmental impact assessment • Three-synchronous policy • Limited time treatment • Centralized pollution control • Two-compliance policy • Discharge permit system (experimental)	• Pollution levy fee • Noncompliance fines • Environmental compensation fee • Sulfur emission fee (experimental) • Emission trading (experimental) • Subsidies for energy-saving products • Regulation on refuse credit to high-polluting firms	• Environmental labeling system • Promotion of ISO 14000 system • Cleaner production	• Cleanup campaign • Performance disclosure

At the beginning of the 1990s, the dominant water pollution issue in China was industrial point-source pollution, specifically pollution from state-owned industries, which became the focus of pollution-control efforts. SEPA and lower level EPBs used a variety of instruments to address the problem, including command-and-control measures, "administrative" measures, economic instruments, and public awareness. The available data suggest that these measures did contribute to reversing industrial water pollution emissions by the middle of the decade.

SEPA, the leading government agency in enforcement, formulates regulations, policies, and standards and uses implementation instruments to prevent and control industrial pollution. The regulatory and administrative framework for pollution control, which today is comprehensive, has been continuously updated and expanded to improve effectiveness and cover emerging issues. During the 1990s, regulators attempted to encourage three shifts in pollution-control strategy, described in the following paragraphs.

First, at the enterprise level, pollution control shifted toward whole-process control, rather than end-of-pipe waste treatment. The shift was encouraged by the promotion of cleaner production concepts and the adoption of ISO 14000 certification procedures. The Cleaner Production Program (CPP), which began in 1993, encourages enterprises to adopt in-plant waste-minimization technologies. Although CPP has great potential, efforts so far have been focused mainly in areas with strong incentives, such as areas of northern China where water is scarce. Nevertheless, the technical capacity to conduct clean-production audits and feasibility studies has been established, and regulators now have the capacity to respond as the demand for these services. In the meantime, the Law of Promoting Cleaner Production, passed and put into effect on January 1, 2003, includes both mandatory and voluntary incentives for enterprises to take measures to control pollution.

ISO 14000 certification procedures were introduced in 1997. The government believes that promoting ISO 14000 will increase the compliance rate without increasing regulatory effort. The project was initiated by SEPA's Office of Environmental Management Systems and was taken over by the Steering Committee for Environmental Management System Certification, which was established within SEPA to provide accreditation services for certification bodies and auditors. Several environmental management and consulting centers have been established to conduct ISO 14000 certification. So far, the majority of participants are either foreign firms or domestic firms engaged in the production of export-oriented products, neither of which is the sort of industry responsible for the nation's main pollution problems. Both CPP and ISO 14000 programs can help reduce China's industrial pollution, but increased funding and manpower will be necessary to enforce pollution laws and regulations.

Second, the regulatory focus has shifted from pollutant concentrations toward total-load control through a combination of concentration and mass-based discharge criteria. In 1987, SEPA (then NEPA) began to experiment with mass-based control to cap or reduce the total level of pollutants released to the environment in certain areas. Seventeen cities and one river basin were selected to participate in the trial implementation of a discharge permit system based on the total-load-control concept. By 1994, the program had expanded to more than 200 cities and 12,000 enterprises. However, studies of the pilot program showed only moderate improvements in environmental quality in participating cities because the program did not address the underlying problem of lax regulatory enforcement.

And third, to improve cost effectiveness, the focus has shifted from dispersed-source control, which requires that each enterprise individually resolve its emission problems, to a more integrated approach with centralized control by, for example, encouraging the discharge of semi-treated liquid wastes to municipal sewers. The previous emphasis on controlling wastes from individual sources did not have the desired results and placed too many demands on EPB staff for supervision; in addition, in less developed areas, there was not enough technical expertise to implement the policy. Encouraging the discharge of pretreated wastes to municipal sewers does not eliminate the need for supervision, but it reduces the monitoring workload and will improve the economics of the centralized sewage system—as long as discharge fees cover the costs of receiving and treating the wastes, and the fees are collected.

Public disclosures of environmental performance by industries and local governments have begun recently. The environmental performance of industries is rated by five color categories (green, blue, yellow, red, and black) from the best to the worst; color ratings are disseminated to the public via the media. Preliminary analyses show that the disclosures have been effective (Wang et al., 2004).

Enforcement

As the highest level of authority, the NPC is responsible for environmental legislation. Supervision and monitoring are mostly the responsibility of law enforcement agencies, including SEPA and local EPBs, which both have departments of environmental supervision and inspection. There are also some supervision and inspection institutes, such as the South China Environmental Protection Supervision and Inspection Center. All of these departments and institutes are responsible for making sure industrial polluters are taking measures to control pollution.

The People's Procurotorates supervise the enforcement agencies to make sure they properly enforce the laws. In addition, the NPC's Environment and Resources Protection Committee organizes yearly nationwide inspections of legal enforcement. The NPC committee also works with the media to organize media coverage of enforcement issues. Media organizations themselves naturally take on the social responsibility of exposing both positive and negative stories in industrial pollution control. The public is also a powerful force for change; as awareness of pollution increases, the public is gradually becoming a nationwide watchdog. Despite all of the progress, however, negotiation and bargaining in environmental policy enforcement are still widespread in China (Wang et al., 2003; Wang and Wheeler, in press).

INSTITUTIONAL PROGRESS

History and Evolution

Progressive institutional reforms since the late 1970s have moved China toward a more decentralized government system. Environmental institutions in China have been evolving along with the reform process. Moves toward more openness, which began in 1978 to bring China out of its isolation from the rest of the world, have led to an era of transition and exploration for a new, workable power structure to govern the country.

Decentralization[3] is essential to the reforms. The government believes that decentralization can cut the central government's costs and improve efficiency by reducing the size of the central bureaucracy. In addition, decentralization is believed to improve the delivery of services by bringing decision making and implementation closer to the target population. Decentralization is also regarded as a natural complement to economic liberalization and the imposition of fiscal discipline.

In the last three decades, major steps toward decentralization to meet the demands of a transition to a market-oriented economy have included: establishment of the Household Responsibility System, which has dismantled the collective ownership of agriculture; the transfer of central authority to local governments and enterprises; the establishment of a socialist market-economy system; and membership in the World Trade Organization. In short, fundamental changes have been made in China's power structure and in the roles of actors in the political, economic, and social arenas.

In 1978, the Household Responsibility System privatized land-use rights in rural areas by leasing agricultural land. Farmers initially received leases on their land for one to three years. In the next decades, the leases were lengthened to 30 and later 50 years. The contract responsibility system was extended to forestlands in the early 1980s and grasslands in the 1990s. Thus, the collective ownership of agriculture was ended, which radically altered the functions of government.

[3]Decentralization refers to any act in which a central government formally cedes powers to actors and institutions at lower levels in a political-administrative hierarchy. It is an umbrella term that includes political decentralization, deconcentration or administrative decentralization, fiscal decentralization, privatization, and even co-management.

In 1982, the government passed legislation that devolved fiscal and economic decision-making responsibilities from the central and provincial levels to the county, township, and even village levels. Under the Fiscal Responsibility System, each level of government became financially independent, that is, responsible for raising and managing its own revenues. (Before 1980, local governments remitted tax revenues directly to the central government and awaited partial returns based on the central government's discretion.) Fiscal decentralization meant new responsibilities for township governments.

In 1988, a new round of government restructuring focused on promoting changes in government responsibilities and functions. Since that time, economic management departments that had directly managed the economy have changed to indirect management to improve administrative efficiency and speed up administrative legislation. Another round of restructuring took place in 1993 and was focused on separating government from business, separating rule-making from implementation, strengthening supervision, and reducing the direct government management of enterprises.

In 1998, the State Council established a government administrative system that would support China's socialist economic system. The government was restructured on the basis of "streamlining, unification, and efficiency," and assignments and divisions of responsibilities were modified in accordance with the principle of "consistency of power and responsibilities." The legal administrative system was strengthened to ensure that the country is ruled by law and governed by law. Similar reforms were carried out at the provincial, municipal, and county levels in the following years. The 1998 reform is regarded as the most dramatic in terms of government restructuring.

The most recent reforms, in 2003, go one step further by instituting a much better developed administrative management system that is standardized, coordinated, transparent, clean, and highly efficient. The most important motivation for this set of reforms was China's accession to the World Trade Organization. A major feature of the 2003 reforms is the separation of decision making, enforcement, and supervisory responsibilities. Along with the internal supervision of administrative departments, the reforms strengthen outside oversight by society, the media, people's congresses, and political consultants.

Compared with many other countries in the world, China has made little progress toward political democratization. Six rounds of government restructuring, however, have led to many achievements. For example, the number of government agencies has declined, and the functions of government agencies at different levels have changed dramatically. The government has gradually given up control over materials allocation, prices, and enterprise operation and management. Today, many state-owned enterprises are characterized by shareholder ownership, market operation, and enterprise-level management. In addition, the private sector has grown rapidly and is now a strong pillar of national economic development. In short, China has moved closer to a market economy.

Dramatic changes have also occurred in the way the government operates. Legal forces are at play in all economic sectors. Rule by law and administration by law have become basic requirements of government operation. In the meantime, government powers continue to devolve, and public participation in government services continues to increase. A transparent government with limited powers, multiple power centers, under the rule of law will provide favorable conditions for China to move further toward political democratization, at least theoretically.

Implications for the Environment

Reforms and decentralization in the last three decades have brought about changes in the way China deals with its daunting environmental challenges. First, responsibilities for the environment and natural resource decision making are being shifted to lower levels of government. Local governments now have not only the authority necessary to reform local industries, but also the power to control financial resources at the local level.

Second, the streamlining of environmental management has become a major feature of the delegation of environmental jurisdiction. Authorities with cross-cutting responsibilities in the environment and sustainable development were greatly strengthened after the 1998 government reforms, when SEPA was promoted to its present ministerial status and SDRC was given a clear mandate to lead China's sustainable development efforts in close collaboration with a number of ministries and agencies.

Third, as the market becomes increasingly decentralized, market forces are gradually assisting command-and-control mechanisms to protect the environment and ecosystem. The government has been experimenting with market-oriented tools in environmental management. Regulators have learned how to price drinking water and utilities based on market values, how to trade SO_2 pollution rights, and how to use taxation to regulate pollution-heavy industries.

And fourth, the concept of environmental governance has become part of the national discourse to encourage information disclosure and public participation. The burgeoning government structure in the environmental arena is reflected by increasing media coverage of environmental issues, and public participation is reflected in the remarkably independent grassroots and citizens' organizations that have been formed in the last several years.

Environmental Institutions

China's environmental awakening happened around 1972 when the first United Nations (UN) Conference on the Human Environment was held in Stockholm, Sweden. Even though at that time China was still a closed society in the middle of the Cultural Revolution, China sent a delegation to the conference.

Thirty years later, the Chinese government has recognized the value of using the market to make environmental protection profitable and to motivate a knowledgeable population to demand compliance and environmental quality.

"Environment" as defined by the Environmental Protection Law of People's Republic of China, refers to the totality of all natural elements and artificially transformed natural elements that affect human existence and development. The environment includes the atmosphere, water, seas, land, minerals, forests, grasslands, wildlife, natural and human remains, nature reserves, historic sites and scenic spots, and urban and rural areas. According to the Environmental Protection Law, environmental responsibility includes protecting and improving both people's living environment and the ecological environment, preventing and controlling pollution and other public hazards, and safeguarding human health.

As a result of the 1972 UN conference, the Chinese delegation realized that China shares the problems of environmental degradation with the rest of the world. The conference was followed a year later by the first National Conference on Environmental Protection in Beijing, and in 1974, by the establishment by the State Council of the Group on Environmental Protection. These events were followed by a series of legislative efforts in the late 1970s, marking the beginning of government efforts to put environmental protection on the national agenda.

In 1984, the group on Environmental Protection was disbanded, and the National Environmental Protection Bureau (NEPB) was set up in the Ministry of Urban and Rural Construction. In 1988, the NEPB was reclassified as an agency, the National Environmental Protection Agency (NEPA), with a bureaucratic rank slightly below a ministry and resumed reporting directly to the State Council. In 1993–1994, the Committee for Natural Resources and Environmental Protection was set up under the NPC to take responsibility for revising and drafting environmental laws, ensuring their rapid promulgation, and supervising their enforcement. And in 1998, NEPA was upgraded to ministerial status and renamed the State Environmental Protection Administration (SEPA).

SEPA underwent dramatic institutional changes in the 1990s. The most notable developments included: (1) elevation to the status of a ministry, clarifying SEPA's position as the agency with overall responsibility for environmental management and protection in China; and (2) efforts to address the so-called "horizontal-vertical" issue, in which lower level EPBs report to higher level EPBs, and ultimately SEPA, but receive their funding from local governments. The heads of local EPBs must now be endorsed by a higher ranked environmental agency.

Lower level EPBs were strengthened, either by raising their bureaucratic status or by giving them independent bureaucratic status. As of 2000, all 31 provincial EPBs were independent agencies, and 30 of them were first-tier institutions; all city-level EPBs were independent agencies, and most were first-tier institutions; about 70 percent of county EPBs were independent; and about

1,422 environmental protection units were established at the township level (World Bank, 2001).[4]

Political Agenda

The Chinese government has clear environmental policy goals that were most recently articulated in former President Jiang Zemin's report to the 16th Communist Party Congress. Sustainable development was identified as one of the three pillars of the strategy to build a "well-off society in an all-round way." An essential element of that strategy is to address "the growing contradiction between the ecological environment and natural resources on the one hand and economic and social development on the other."

Over the last two decades, the environment has been given increasing priority. In the Tenth Five-Year Plan (2001–2005), sustainable development is designated as the "guiding principle and strategy" for the country's economic and social development. The Five-Year Plan also identifies environmental protection as a national priority, but not quite on the same level as economic growth or the alleviation of poverty. This has been one of the major reasons some environmental regulations have not been effectively enforced.

Legal System

The effectiveness of the environmental legal system, which is at an early stage of development, is hampered by a variety of factors. There is a deeply ingrained problem-solving culture on the part of all parties—EPBs, other government agencies, and industrial enterprises—based on negotiating and bargaining outside of the court system. This tradition militates against turning to the courts for recourse and will have to be overcome. This is compounded by the lack of a tradition of drafting laws to limit ambiguity and define the rights and responsibilities of all relevant parties.

To strengthen environmental law enforcement, the government has revised criminal laws a few times to include punishment for violators of laws and regulations to protect the environment and resources and regulations of those causing damage to the environment, property, or public health. The current Section 6 of Chapter VI specifically designated "Crimes of Undermining Protection of Environmental Resources," specifies three- to seven-year prison terms and monetary penalties for violations of the legal restrictions on the discharge, dumping, or disposal of radioactive waste, hazardous waste, and other dangerous substances to the environment. Violators of laws or regulations on protecting aquatic

---

[4]The World Bank, China Air, Land and Water: Environmental Priorities for A New Millennium, August 2001, pp. 99-100.

resources, the hunting and killing of protected species, the occupation of or serious damage to forestland or cultivated land, logging and mining, etc., are also subject to punishment.

Although law enforcement agencies have the power to enforce environmental laws, the facts show that a tremendous burden still falls on the shoulders of victims, who must collect proof to make a case against violators. So far, the enforcement of environmental laws and regulations has been spotty.

Interagency Coordination

With sustainable development and the environment high on the national agenda, one would expect that a mechanism would be in place for integrating and coordinating environmental and economic policies. However, this is not the case. The State Council used to have an Environmental Protection Commission, set up in 1984, to aid in cross-sectoral coordination. The commission was disbanded in 1998 when SEPA was elevated to the ministerial level and the commission's coordinating role was transferred to SEPA. With the dissolution of the commission, China lost a potentially important forum for encouraging collaborative, cross-cutting approaches to environmental issues and for resolving jurisdictional disputes. Experience in other countries suggests that SEPA, a second-rank (non-cabinet) ministry, will not be able to carry out this role effectively because it is below the ministerial level in many respects and because it is considered somewhat of a newcomer. Moreover, many enterprises, local governments, and production-oriented ministries view SEPA as an environmental policeman trying to limit economic growth. In response to these reactions, SEPA tends to adopt a defensive attitude and stick to the issues it feels most comfortable with—issues that can be addressed without collaboration with other agencies.

On paper, all major economic development policies in China must go through an environmental impact assessment to ensure that they do not entail negative impacts on the environment. But in reality, the questions of who should perform the assessment, how it should be done, how the assessment will impact economic policies, and so forth, have not been answered.

The lack of coordination can be illustrated by the example of ecological conservation. This is an area in which many government agencies and departments have responsibilities (e.g., SEPA, SFA, MOA, MOWR) and even conflicting interests. Without the State Council's Environmental Protection Commission or a similar group, there is no linkage or connection among the plans made by those agencies. In addition, some agencies have internally conflicting mandates; SFA, for example, is in charge of both the conservation and exploitation of forests.

Contractual Arrangements

SEPA holds local decision makers accountable for the environmental quality in their localities. The performance of governors and mayors is evaluated partly by the environmental quality and improvements in their jurisdictions. With this kind of contractual power, SEPA should be able to ensure that many environmental policies are implemented at the local level. However, no penalty mechanisms have been designed or implemented for contracts that are not fulfilled. However, public disclosures of local government environmental performance can provide incentives for local governments to enforce regulations enacted by the central government.

Nongovernmental Organizations and Research Institutes

Authoritative advice from independent, reliable sources is crucial to sound decision making, especially for the complex issues involved in sustainable development. In the past, technical and academic institutes were part of, or under the supervision of, government ministries or agencies for which they gathered information. Not surprisingly, these institutes provided advice that was carefully attuned to "the master's voice" (any notion of potentially adverse effects was routinely brushed under the carpet).

As a result of recent government restructuring and reform, many of these organizations are now more independent. Combined with their improved capacities, they are now in a position to give objective, even challenging advice to decision makers. In addition, there is a growing, competitive market for environmental knowledge and services, mostly government agencies contracting for decision support, such as studies, investigations, and feasibility studies. Finally, increasing pluralism in the technical and academic sectors has created more opportunities for institutes to network and collaborate with research institutes and consulting agencies around the world; Chinese institutes have begun bidding for overseas consulting jobs, as well as international research assignments. The major concern is that NGOs and think tanks are not allowed to operate independent of official government policy.

Economic Globalization and International Cooperation

Trade and commerce are also general stimuli to environmental protection. With globalization and opening markets, China is interacting more and more with its neighbors and with the rest of the world. Trade and commerce are growing at an unprecedented rate. All of these interactions are driving environmental and social change in China, and laws and regulations have been developed specifically to ensure environmental protection in imports and exports and general trade (see Tables 4, 5, and 6).

TABLE 4 Import and Export of Environment-Related Goods

Items Specifically Listed	Agency
Prohibited imported goods	Ministry of Foreign Trade and Economic Cooperation (MOFTEC)
Wastes used as raw materials and restricted in importation	MOFTEC, General Administration of Customs, State Administration of Quality Supervision, Inspection and Quarantine, SEPA
Waste used as raw materials under automatic import license category	SEPA, MOFTEC, etc.
Machinery and electronic products subject to automatic import license	MOFTEC, etc.
Goods subject to import license	MOFTEC, etc.
Goods for which quota, license and specific import administration measure are eliminated	MOFTEC
Prohibited exported goods	MOFTEC
Prohibiting export of black moss	MOFTEC

TABLE 5 Import and Export of Hazardous Chemicals

Topics	Agency
Circular on strengthening the management of hazardous chemicals	SEPA, MOFTEC, Ministry of Public Security, etc.
Provisions on environmental management of the first import of chemicals and the import and export of toxic chemicals	SEPA, MOFTEC
Standards of registration fee for environmental management of chemical export and import	SEPA
Circular on renewal of registration of environmental management of first import of chemicals	SEPA

Through environment-related provisions in the legal agreements for China's accession into the WTO, China is obligated to abide by various requirements. The goal of WTO is to establish global international competition under equal conditions. To be competitive in this market, Chinese industry must become more efficient, which, at least in the long run, may lead to better use of resources, better conservation, and the importation of cleaner, more efficient technology.

TABLE 6 Trade Provisions in Multilateral Environmental Agreements

Basel Convention on the Control of Transboundary Movement of Hazardous Wastes and Their Disposal
Amendment to the Basel Convention
Rotterdam Convention on the Prior Informed Consent Procedure for Certain Hazardous Chemicals and Pesticides in International Trade
Stockholm Convention on Persistent Organic Pollutants
Framework Convention on Climate Change
Kyoto Protocol to the UN Framework Convention on Climate Change
Montreal Protocol on Substances that Deplete the Ozone Layer
Convention on Biological Diversity
Cartagena Protocol on Biosafety
Convention on International Trade in Endangered Species of Wild Fauna and Flora
Agreement on Fish Stocks
International Tropical Timber Agreement
Convention on the Ownership of Cultural Property
Rio Declaration on Environment and Development
Agenda 21
Statement of Principles on Forests

The economic activities of China and Chinese enterprises have an impact on both regional and global environments. Although no detailed studies of China's ecological footprint have been undertaken, there are five relevant areas for study: (1) greenhouse gas emissions; (2) biodiversity; (3) transboundary pollution; (4) domestic consumption; and (5) overseas investment. For instance, severe dust storms, resulting in large part from desertification throughout China, have played havoc with air quality and transportation in China and neighboring countries, such as Korea and Japan. Dust plumes from these storms have even been identified in the United States, reportedly transported via the jet stream.

One positive aspect of regional and global environmental developments is that they have led to meetings between high-level government officials in the affected countries to discuss solutions to China's growing environmental problems. In April 2002, the environmental ministers of China, South Korea, and Japan met in Seoul for the 4th Tripartite Environment Ministers Meeting. Most of the discussions in 2002 were focused on the problems of "yellow dust" emanating from China. South Korea initiated the meeting in 1999 to address transboundary pollution originating in China. Since then, the meetings have addressed many environmental issues of mutual concern.

China has worked with other countries bilaterally and multilaterally to protect the environment in border areas. For example, the Mekong River Watershed Management Committee is a regional organization with members from Southeast Asia along the Mekong River. Although China has not become an official member of the committee, China is involved along the upper reaches of the Mekong, which originates in the Tibet-Qinghai Plateau. Thus, what China does along the watershed and how China plans to use the water resources of the upper reaches of the Mekong have a major effect on the river itself and the development of countries along the middle and lower reaches of the river. The committee brings the stakeholders together to discuss how the watershed can be protected while economic growth is pursued.

Other examples include the China-Russia collaborative program to protect the Usuri River and China/Myanmar/Laos cooperative efforts to control trafficking in endangered species. However, projects like these have many problems, such as lack of financial support, weak enforcement powers, and corruption in many countries.

THE FUTURE OF POLLUTION CONTROL IN CHINA

There is no doubt that China faces major environmental challenges. If GDP continues to grow at (or near) a rate of 8 percent in the short term or medium term, urbanization continues and possibly even accelerates, and industrialization continues to evolve toward the production of finished products, the scope and dimensions of China's environmental problems will increase and become more complex. Therefore, China must continue to improve its environmental institutions.

The decentralization reform process has generally had a positive impact on the environment. Despite uneven progress in economic, public administration, and governance reforms, there have been large environmental payoffs. The greatest impact on the environment has been the result of a more efficient economy, which has included more efficient use of resources, decreased industrial and domestic wastes, and a system of incentives that encourages further environmental improvements.

However, the reactive approaches of the past will not be sufficient to address the challenges China faces. The government must become more proactive in addressing environmental problems. Given the substantial and growing tension between economic development and environmental protection, the government must ensure that environmental factors are considered in policy decisions, and there must be comprehensive environmental supervision of all government activities. And government should maintain its strong commitment to the environmental agenda, which includes improving the quality of life rather than promoting economic development at the expense of the environment.

Continued reform of public administration toward good governance for sustainable development is essential to stimulating economic reform and a necessary basis for reinforcing the environmental legislative framework and ensuring the implementation of environmental laws and regulations. The growing cadre of more efficient public officials at the national level and across China, with increasingly clear duties and responsibilities and a commitment to the rule of law, bodes well for the protection of the environment. Further clarifications and separations of responsibilities, from national to local levels and among agencies at the same level, will yield similarly impressive results.

Efforts to strengthen SEPA and EPBs should be continued, and SEPA should be elevated to membership in the State Council to ensure that the government's environmental concerns are taken into account in all aspects of development policy. SEPA's institutional capacity, including staffing levels, should be significantly increased to bring it more into line with comparable international agencies. The lack of cross-sectoral coordination is commonly considered to be at the heart of the critical problems related to water resources management, urban planning, forest management, product-related environmental measures, and many other areas. To improve cross-sectoral coordination, a national environmental network, with SEPA as its anchor, should be developed; in addition, SEPA's environmental coordination function should be enhanced.

In the meantime, limited experiments in governance reforms have shown that empowering and liberalizing people, within a regulated and monitored framework, can produce a strong, positive force for environmental change that complements, rather than threatens, government objectives and actions. Experience in China and other countries suggests that the media and NGOs will be strong advocates for the environment in the future. A law ensuring the disclosure of environmental information and public participation in environmental management decisions should be enacted, and the development of environmental NGOs should be promoted.

BIBLIOGRAPHY AND REFERENCES

Dupar, M., and N. Badenoch. 2002. Environment, Livelihoods, and Local Institutions: Decentralization in Mainland Southeast Asia. Washington, D.C.: World Resources Institute.

Fengzhong, C., D. Guo, G. Zhou, and H. Wang. 2002. Industrial Pollution Control Policies in China: Evaluations and Recommendations. A working paper for the World Bank, April 2002. Washington, D.C.: The World Bank.

Jahiel, A.R. 1998. The organization of environmental protection in China. China Quarterly 156(Special Issue on China's Environment): 757–787.

Qiao, M., ed. 2003. Interpreting the New Government: Policy Direction of the Crucial Issues Related to National Economy and the People's Livelihood. Beijing: Chinese Communist Party History Press.

SEPA (State Environmental Protection Administration). 2003. Environmental Laws and Regulations in China and Rules of WTO, Policy and Laws. Beijing: Guangming Daily Press.

SEPA. 2003. State of the Environment Report, 2003. Beijing: State Environmental Protection Administration.

Stockholm Environment Institute and the United Nations Development Program. 2002. Making Green Development a Choice. China Human Development Report. Oxford, U.K.: Oxford University Press.

Wang, H., and D. Wheeler. In press. Financial incentives and endogenous enforcement of China's pollution levy system. Journal of Environmental Economics and Management.

Wang, H., and C. Wu. 2004. Public Access to Environmental Information in China. Washington, D.C.: World Bank Development Research Group.

Wang, H., N. Manmingi, B. Laplante, and S. Dasgupta. 2003. Incomplete enforcement of pollution regulation: bargaining power of Chinese factories. Environmental and Resource Economics 24(3): 245–262.

Wang, H., J. Bi, D. Wheeler, J. Wang, D. Cao, G. Lu, and Y. Wang. 2004. Environmental performance rating and disclosure: China's GreenWatch program. Journal of Environmental Management 71(2): 123–133.

World Bank. 2001. China Air, Land, and Water: Environmental Priorities for a New Millennium. Washington, D.C.: The World Bank.

Wu, F. 2002. New Partners or Old Brothers?: GONGOs in Transnational Environmental Advocacy in China. China Environment Series 5. Washington, D.C.: Woodrow Wilson International Center for Scholars.

Xu, J., E. Katsigris, and T.A. White, eds. 2002. Implementing the Natural Forest Protection Program and the Sloping Land Conversion Program: Lessons and Policy Recommendations. Beijing: China Council for International Cooperation on Environment and Development. Also available online at: *http://www.harbour.sfu.ca/dlam/Taskforce/grassfindingindex.html.*

Zhao, L. 1998. Reform of Government Administration: China Perspectives toward the 21st Century. Jinan, China: Shandong People's Press.

Public Health

Ambient Air Pollution in Shanghai: A Health-Based Assessment

HAIDONG KAN and BINGHENG CHEN
Department of Environmental Health
School of Public Health
Fudan University

CHANGHONG CHEN
Shanghai Academy of Environmental Science

To investigate the potential impact of ambient air pollution on public health and the economy in Shanghai, we estimated exposure levels of the general population under various planned energy scenarios and then assessed potential health impacts using concentration-response functions derived from available epidemiologic studies. We then estimated the corresponding economic values of the health effects based on the unit values of the health outcomes. The results showed that ambient air pollution could have a significant impact on the health of Shanghai residents, both in physical and monetary terms. Compared with the base-case (BC) scenario, implementation of various planned energy scenarios could prevent many premature deaths (from 608 to 5,144 in 2010 and from 1,189 to10,462 in 2020) and could substantially reduce the number of pollution-related diseases. The expected economic benefits vary widely depending on the underlying assumptions. Thus, energy and environmental policy could not only reduce air pollution and improve air quality, but could also protect public health and benefit the economy.

Energy and related health issues are of growing concern worldwide. Fossil fuels, the primary sources of energy, are also the greatest sources of ambient air pollution—nitrogen oxides (NO_x), sulfur dioxide (SO_2), dust, soot, smoke, and other suspended fine particulate matter (PM). These pollutants can lead to serious public health problems, including asthma, irritation of the lungs, bronchitis, pneumonia, decreased resistance to respiratory infections, and premature death. The burning of fossil fuels is also the major source of carbon dioxide (CO_2) emissions, a primary contributor to global warming.

In China, a developing country, coal has been the primary energy source.

Rapid economic growth has been accompanied by a rapid increase in the demand for energy, and the relatively inefficient energy technology currently used in China has caused high pollution emissions. However, China has been making great efforts to save energy, optimize its energy structure, and increase energy efficiency to balance energy consumption and environmental needs. As a result of these efforts, the increase in energy consumption has been lower than the increase in gross domestic product (GDP) in China during the past decade.

Clearly, energy-policy decisions today will have a significant impact on future air pollution levels and public health. In the present study, we estimate the public health impact of ambient air pollution under various energy scenarios in Shanghai, one of the fastest growing urban areas in China, and place monetary values on the estimated health effects.

METHODS

Development of Energy Scenarios

The MARKAL (MARKet ALlocation) optimization model was used to estimate pollutant emissions under various energy scenarios in Shanghai. MARKAL is a dynamic, linear programming model that optimizes a technology-rich network representation of an energy system. MARKAL models the economy of a region as a system, represented by processes and physical and monetary flows among those processes. Details about the application of MARKAL for energy and environmental policies in Shanghai have been discussed elsewhere (Gielen and Chen, 2001).

The energy scenarios developed for Shanghai can be classified into four

TABLE 1 Elements of the Scenarios

	Economic Growth	Improvement in Energy Efficiency	Maximum Coal Use	Electricity Imports
BC	√			
EFF	√	√		
NG	√	√	√	√
SO_2	√	√	√	√
NO_x	√	√	√	√
CO_2	√	√	√	√

categories: (1) base-case (BC); (2) energy options (i.e., use of efficiency measures [eff] or natural gas [NG]); (3) pollution targets (i.e., constraints on SO_2 or NO_x); and (4) control of CO_2 emissions. The details of the scenarios are shown in Table 1.

Concentrations of Ambient Air Pollutants

Based on the principle of transfer matrix, an air-quality model (the Exposure Level Model) was developed to link MARKAL emission scenarios and concentrations of ambient air pollutants. Input to the fundamental matrix was a long-range transport and deposition model (ATMOS model) for SO_2 and primary PM_{10} (particulates with diameters of 10 mm or less). The ATMOS model is a Lagrangian parcel model with three vertical layers (Calori and Carmichael, 1999).

For the Shanghai project, the ATMOS model provided a 4×4 km resolution of the concentration of SO_2 and primary PM_{10}. The total area of Greater Shanghai, 6341 km^2, was divided into 487 grids. Based on matrix output of the ATMOS model, the Shanghai Exposure Level model was developed in Excel to link MARKAL predictions and provide exposure levels for the analysis of health impacts.

Exposure Levels

Because strong epidemiologic evidence supports the association of PM_{10} with adverse health effects, PM_{10} was selected as a useful indicator of several sources of outdoor air pollution, such as fossil-fuel combustion (Wilson and Spengler, 1996).

Availability of Natural Gas	Control of SO_2 Emissions	Control of NO_x Emissions	Control of PM_{10} Emissions	Tax on CO_2
√				
√	√			
√	√	√		
√	√	√	√	√

Everyone living in Greater Shanghai was considered to be in the exposed population in this analysis. The number of Shanghai residents in each 4 × 4 km grid cell was estimated based on population data from the Shanghai Bureau of Statistics. Combining the PM_{10} level and population in each grid cell, we estimated exposure levels to outdoor air pollution under various scenarios in 2010 and 2020.

Estimations of Health Effects

To develop estimates of public health impacts of air pollution, we used concentration-response (C-R) coefficients derived from published, peer-reviewed studies in China and worldwide.

Because most epidemiologic studies linking air pollution and health effects are based on a relative-risk model in the form of a Poisson regression, health effects at a given concentration, *C*, could be given by the following equation:

$$E = exp(\beta(C-C_0)] \qquad (1)$$

C and C_0 are the PM_{10} concentrations, and E and E_0 are the corresponding health effects under a specific scenario and the BC scenario, respectively. β is the exposure-response function.

Exposure-response functions (β) link air quality changes and heath outcomes. Our preference was to use C-R functions from Chinese studies whenever they were available. Results in the international peer-reviewed literature were used only when the selected end points could not be found in the Chinese literature. If there were several studies describing a C-R function for the same health end point, we used the pooled estimate to get the mean and 95 percent confidence interval (*CI*) of the coefficient.

The results of the analysis were given as a comparison of the health effects under a specific scenario with the health effects under the BC scenario in 2010 and 2020, respectively.

Economic Value of Health Effects

The analysis was based mainly on the concept of willingness to pay (WTP); the cost of illness (COI) was used as an alternative for morbidity end points that could not be valued based on the existing literature.

The effect of air pollution on mortality was assessed by using the *value of a statistical life* (VOSL). The literature on VOSL and WTP to avoid a statistical premature death is mainly from the United States. However, because of our limited time and budget, we used a contingent valuation study (CVM) conducted in Chongqing, China, (Wang, 2001) for the estimate of Shanghai VOSL. In the Chongqing study, Wang reported an average WTP of US$34,750. The marginal effect of increased income on WTP was also reported: with an increase in annual income of US$145.80, the increase in WTP was US$14,550. Taking the annual income difference in 2001 between Chongqing (US$495.7) and Shanghai (US$1,234.5) into account, we did a conversion based on Chongqing's coefficient between marginal WTP and income to get an estimate for the VOSL in Shanghai.

Because data for China were not available for some end points of morbidity, we used a value based on the conversion used by the U.S. Environmental Protection Agency. The ratio for conversion was based on the per capita income of U.S. and Shanghai residents, and the income elasticity was assumed to be 1. The COI for the end point of hospital admissions and outpatient visits was calculated from data from Shanghai because no studies based on WTP were available for those end points.

The economic value of a change in the incidence of an adverse health outcome was calculated as the change in incidence (the number of avoidable deaths) multiplied by the unit monetary value (the value of a single case avoided). Because of inherent uncertainties in the health and economic impact assessment, the results were given as a range (mean and 95 percent CI). Because both health outcomes and unit values were distributions rather than constants, we conducted a Monte Carlo simulation to calculate the economic values.

RESULTS

Exposures to Particulates in Shanghai

Table 2 summarizes the percentage of the population exposed to different levels of PM_{10} under various scenarios in 2010 and 2020.

It should be emphasized that the PM_{10} levels in Table 2 are much lower than the actual concentrations in Shanghai because only PM_{10} from energy consumption was included. PM_{10} from other sources, such as natural sources, construction sites, and so forth, was not included.

Estimates of Health Effects

Table 3 summarizes the PM_{10} exposure-response coefficients (mean and 95 percent CI) and baseline rates of selected health outcomes. The excess cases in each scenario (compared to the BC scenario) were computed based on the change in exposure levels to PM_{10} under each scenario, exposure-response functions, and baseline rates for the health outcomes.

The potential health benefits in 2010 and 2020 shown in Table 4 clearly indicate that the choice of energy scenario could have a significant impact on the health of Shanghai residents. Compared with the BC scenario, the implementation of other energy scenarios could prevent 608 to 5,144 PM_{10}-related deaths in 2010 and 1,189 to 10,462 PM_{10}-related deaths in 2020.

TABLE 2 Exposure Levels of PM_{10} under Different Scenarios for 2010 and 2020

	2010					
PM_{10} level ($\mu g/m^3$)	BC	EFF	NG	SO_2	NO_x	CO_2
<5	–	–	0.1	0.1	0.4	0.8
10–15	2.0	2.3	4.5	4.7	7.5	9.6
15–20	6.4	6.8	9.2	9.8	12.5	13.3
20–25	8.2	8.5	11.1	10.8	8.3	7.9
25–30	8.3	8.0	5.1	5.5	6.1	8.2
30–35	4.4	4.2	5.3	5.8	13.0	44.6
35–40	3.3	4.5	9.1	12.6	51.6	15.6
40–45	4.8	5.2	35.7	38.2	0.6	–
45–50	7.0	11.1	19.9	12.5	–	–
50–55	18.1	31.5	–	–	–	–
55–60	–	–	–	–	–	–
60–65	–	–	–	–	–	–
65–70	–	–	–	–	–	–
70–75	–	–	–	–	–	–
>75	–	–	–	–	–	–

Economic Value of Estimated Health Effects

Table 5 summarizes the unit values (mean and 95 percent CI) for various end points in 2000 in Shanghai and the specific approach used to derive them. By combining the health benefits and unit values, we computed the economic benefits compared to the BC scenario. Table 6 shows the results for 2010 and 2020.

DISCUSSION

The link between energy and health must be taken into account in decisions addressing rapid economic growth and sustainable development in Shanghai. This study, which is based on the same approaches used internationally for assessments of environmental impacts, shows that an effective energy and environmental policy will be an important factor in reducing air pollution, improving air quality, and promoting public health.

The quantification of the impact of air pollution on public health is a critical component in environmental policy decisions. Given the gaps in scientific knowledge about the health effects of air pollution and the wide range of uncertainties characterizing many aspects of the process, analyzing the total burden of ambient air pollution on public health is a challenging task.

2020					
BC	EFF	NG	SO_2	NO_x	CO_2
–	–	–	–	0.3	0.9
0.3	0.5	2.9	2.9	6.9	9.6
2.6	2.9	7.6	7.6	12.8	13.8
5.1	5.6	9.9	10.0	7.9	7.6
5.7	6.2	6.5	6.5	5.7	10.6
6.8	6.4	4.6	4.7	12.1	50.1
5.2	5.4	4.2	5.0	44.4	7.4
3.0	2.7	8.4	7.5	9.9	–
2.9	2.9	18.8	19.8	–	–
2.4	3.3	37.1	36.0	–	–
4.0	3.9	–	–	–	–
2.4	8.9	–	–	–	–
12.4	16.0	–	–	–	–
13.3	35.3	–	–	–	–
33.9	–	–	–	–	–

Our estimates were conservative for three reasons. First, in the present analysis we selected only PM_{10} as an indicator of outdoor air pollution and probably overlooked adverse health effects from exposure to other air pollutants; thus, we probably underestimated the health effects attributable to total air pollution. Although PM_{10} may be a good indicator, there is clear evidence that other pollutants, such as ozone, NO_x, and SO_2, have independent health effects. In addition, we did not include synergistic effects among air pollutants or with cofactors, such as pollen and other allergens.

Second, the ATMOS model we used could only deal with primary PM_{10} and SO_2. Thus, we underestimated the health effects attributable to secondary PM_{10}, such as sulfate and nitrate. Previous studies have shown that ammonium sulfate and nitrate account for substantial proportions of fine particles in Shanghai (Ye et al., 2003).

Third, we focused only on health outcomes that could be quantitatively estimated and then translated into monetary values for further assessment. Some end points, such as subclinical symptoms and decreased pulmonary function, were not included in this analysis, although there is evidence of an association between them and exposure to air pollution. We did not estimate cancer-related

TABLE 3 Exposure-Response Coefficients and Baseline Rate for Exposure to PM_{10} (per person)

Health Outcome (by age group)	Mean (95 percent CI)	Reference	Frequency	Reference
Long-term mortality (adult ≥ 30)	0.00430 (0.00260–0.00610)	Dockery et al., 1993; Pope et al., 1995	0.01077	Shanghai Municipal Bureau of Public Health, 2000
Chronic bronchitis (all ages)	0.00450 (0.00127–0.00773)	Ma and Hong, 1992; Jin et al., 2000	0.01390	China Ministry of Health, 1998
Respiratory hospital admission (all ages)	0.00130 (0.00010–0.00250)	Zmirou et al., 1998; Wordley et al., 1997; Prescott et al., 1998	0.01240	Shanghai Municipal Bureau of Public Health, 2000
Cardiovascular hospital admission (all ages)	0.00130 (0.00070–0.00190)	Wordley et al., 1997; Prescott et al., 1998	0.00850	Shanghai Municipal Bureau of Public Health, 2000
Outpatient visits—internal medicine (all ages)	0.00034 (0.00019–0.00049)	Xu et al., 1995	3.26000	Shanghai Municipal Bureau of Public Health, 2000
Outpatient visits—pediatrics (all ages)	0.00039 (0.00014–0.00064)	Xu et al., 1995	0.30000	Shanghai Municipal Bureau of Public Health, 2000
Acute bronchitis (all ages)	0.00550 (0.00189–0.00911)	Jin et al., 2000	0.39000	Wang et al., 1994
Asthma attack (children < 15 years)	0.00440 (0.00270–0.00620)	Roemer et al., 1993; Segala et al., 1998; Gielen et al., 1997	0.06930	Ling et al., 1996
Asthma attack (adults ≥ 15 years)	0.00390 (0.00190–0.00590)	Dusseldorp et al., 1995; Hiltermann et al., 1998; Neukirch et al., 1998	0.05610	Ling et al., 1996

effects linked to exposures to ambient air pollution, although recent studies in the United States have suggested an association (e.g., Pope et al., 2002).

Because some of the exposure-response functions we used in this analysis were not available in Chinese studies, we relied on international studies, conducted mostly in the United States and Western Europe. This raises the question of whether results from a developed country can be transferred to a developing country. For example, Chinese studies generally report lower coefficients for exposure-response relationships between air pollution and adverse health effects than studies in the United States and Europe. This is probably because of differences in levels of air pollution, local population sensitivity, age distribution, and especially air pollutant components. For instance, the composition of the motor vehicle fleet in Western Europe and the United States, where most of the epidemiological studies were performed, and the motor vehicle fleet in China differs substantially. Another major difference is the widespread use of coal in China, which suggests that the air pollution mix also differs substantially.

Ideally, when exposure-response functions from developed countries are applied to other regions, for example, Shanghai, they should be revised to account for local conditions, such as the physical (diameter, etc.) and chemical (components) properties of particulates, the socioeconomic status of local populations, etc. However, no reference data are available for such revisions. Until locally derived exposure-response functions become available, this will probably be the weakest part of an analysis.

Because no valuation study of the health end points associated with air pollution in Shanghai had been done before, we had to estimate values from previous studies of similar changes, a procedure called "benefit transfer" or "value transfer" in economics. Characteristics of the concerned population (e.g., age distribution, income, health status, culture) may have contextual effects on the valuation results. For example, different social and health-insurance systems greatly influence people's risk perception, which affects the WTP to avoid the risk.

If we had transferred the U.S. VOSL directly to the Shanghai study, after accounting for the income difference between the two sites, the VOSL would have been US$780,000, which is much higher than the VOSL estimated in the Chongqing study. The number would be even higher if we had used purchasing power parity (PPP) as the income definition. Obviously, the estimate used in the Chongqing study is better fitted to the Shanghai study in terms of the economic and social situation. Therefore, we tried to use Chinese studies wherever they were available and attempted to stay on the conservative side with a range of reasonable estimates.

There are also inherent uncertainties in transferring values from other study sites, whether in China or elsewhere. Therefore, we strongly suggest that a WTP study for the avoidance of air-pollution-related health risks in Shanghai be undertaken, especially on the WTP to reduce the risk of premature death from air

TABLE 4 Health Benefits of Different Energy Scenarios Compared with the BC Scenario for 2010 and 2020 (mean value)

	2010				
	EFF	NG	SO_2	NO_x	CO_2
Premature death	608	2,761	3,079	4,266	5,144
Chronic bronchitis	1,315	5,964	6,649	9,210	11,100
Respiratory hospital admissions	377	1,740	1,943	2,712	3,286
Cardiovascular hospital admissions	260	1,197	1,336	1,865	2,260
Outpatient visits (internal medicine)	27,080	125,400	14,0200	196,000	237,900
Outpatient visits (pediatrics)	2,807	13,000	14,530	20,320	24,660
Acute bronchitis	49,490	223,500	249,000	344,200	414,200
Asthma attacks	1,508	6,858	7,649	10,610	12,790

TABLE 5 Unit Values for End Points in 2000 (in 2000 US$)

End Point	Mean (95 percent CI)	Approach
Premature death	108,500 (101,900–115,100)	WTP
Chronic bronchitis	6,050 (807–20,130)	WTP
Respiratory hospital admissions	710[a]	COI
Cardiovascular hospital admissions	1,043[a]	COI
Outpatient visits (internal medicine)	14[a]	COI
Outpatient visits (pediatrics)	14[a]	COI
Acute bronchitis	7.2 (2.6–11.9)	WTP
Asthma attacks	5.3 (2.3–8.3)	WTP

[a]The available data did not provide a distribution of values.

pollution. Additional research will be necessary to improve our evaluations of the health outcomes associated with air pollution, which should include: factors that influence WTP (e.g., age, income, education level, pollution level); the relationship between WTP and quality of life; the relationship between private costs and lost output; people's preferences in trade-offs among risks.

In summary, despite the limitations of our analysis, the results highlight the importance of considering air-pollution-related health effects in decisions about energy options in Shanghai. As we move toward sustainable development and

2020				
EFF	NG	SO_2	NO_x	CO_2
1,189	6,424	6,541	9,219	10,462
2,580	13,910	14,160	19,940	22,620
704	3,918	3,991	5,707	6,522
485	2,694	2,744	3,924	4,485
49,850	279,600	284,900	409,200	468,600
5,173	29,000	29,540	42,430	48,590
98,520	526,400	535,900	751,300	850,800
2,937	15,910	16,200	22,860	25,960

health, close collaboration between public health policy and energy policy will increase the chances of preventing avoidable health hazards.

The approaches recommended in this analysis can be used to evaluate other regions in China for local and nationwide air-pollution-related health-risk assessments and economic evaluations. But health-impact assessment methods must be improved and refined, especially in dealing with uncertainties, transferring exposure-response functions, and using more common health indicators, such as disability-adjusted life years (DALYs). These improvements will require close cooperation among air-pollution modelers, epidemiologists, economists, and policy makers.

ACKNOWLEDGEMENT

This project was cofunded by the U.S. Environmental Protection Agency (EPA) / U.S. National Renewable Energy Laboratory (NREL) Integrated Environmental Strategies Program (IES), China Council for International Cooperation on Environment and Development, China State Environmental Protection Bureau, China Ministry of Health, and Shanghai Environmental Protection Bureau.

TABLE 6 Economic Benefits of the BC Scenario for 2010 and 2020 (in millions of 2000 US$) (mean value)

	2010				
	EFF	NG	SO_2	NO_x	CO_2
Premature death	104.0	469.0	524.0	729.2	873.2
Chronic bronchitis	7.5	34.0	36.9	52.4	60.7
Respiratory hospital admissions	0.4	1.8	2.0	2.9	3.5
Cardiovascular hospital admissions	0.4	1.8	2.1	2.9	3.5
Outpatient visits (internal medicine)	0.6	2.6	2.9	4.1	4.9
Outpatient visits (pediatrics)	0.1	0.3	0.3	0.4	0.5
Acute bronchitis	0.5	2.2	2.5	3.5	4.1
Asthma attacks	0.0	0.1	0.1	0.1	0.1
Total	113.5	511.8	570.8	795.5	950.5

BIBLIOGRAPHY AND REFERENCES

Calori, G., and G.R. Carmichael. 1999. An urban trajectory model for sulfur in Asian megacities: model concepts and preliminary application. Atmospheric Environment 33(19): 3109–3117.

China Ministry of Health. 1998. Second National Health Service Survey. Beijing: China Ministry of Health.

Dockery, D.W., C.A. Pope, X. Xu, J.D. Spengler, J.H. Ware, M.E. Fay, B.G. Ferris, Jr., and F.E. Speizer. 1993. An association between air pollution and mortality in six U.S. cities. New England Journal of Medicine 329(24): 1753–1759.

Dusseldorp, A., H. Kruize, B. Brunekreef, P. Hofschreuder, G. de Meer, and A.B. van Oudvorst. 1995. Association of PM_{10} and airborne iron with respiratory health of adults living near a steel factory. American Journal of Respiratory and Critical Care Medicine 152(6Pt1): 1932–1939.

Gielen, D., and C. Chen. 2001. The CO_2 emission reduction benefits of Chinese energy policies and environmental policies: a case study for Shanghai, from 1995–2020. Ecological Economics 39(2): 257–270.

Gielen, M.H., S.C. van der Zee, J.H. van Wijnen, C.J. van Steen, and B. Brunekreef. 1997. Acute effects of summer air pollution on respiratory health of asthmatic children. American Journal of Respiratory and Critical Care Medicine 155(6): 2105–2108.

Hiltermann, T.J., J. Stolk, S.C. van der Zee, B. Brunekreef, C.R. de Bruijne, P.H. Fischer, C.B. Ameling, P.J. Sterk, P.S. Hiemstra, and L. van Bree. 1998. Asthma severity and susceptibility to air pollution. European Respiratory Journal 11(3): 686–693.

Jin, L.B., Y. Qin, and Z. Xu. 2000. Relationship between air pollution and acute and chronic respiratory disease in Benxi. Journal of Environment and Health 17: 268–270. In Chinese.

Ling, S., J. Peng, and Z. Wang. 1996. A sample survey and multiple factor analysis on asthma in urban districts of Shanghai. Chinese Journal of Tubercular and Respiratory Disease 19: 11–13.

		2020		
EFF	NG	SO_2	NO_x	CO_2
300.4	1618.4	1647.6	2340.0	2646.0
21.6	117.3	118.5	161.8	188.8
1.1	6.1	6.2	8.9	10.2
1.1	6.2	6.3	9.0	10.3
1.5	8.6	8.7	12.6	14.4
0.2	0.9	0.9	1.3	1.5
1.5	7.8	7.8	11.1	12.8
0.0	0.2	0.2	0.3	0.3
327.4	1765.5	1796.2	2544.5	2884.3

Ma, H.B., and C.J. Hong. 1992. Effects of particulate air pollution on respiratory disease. Chinese Journal of Public Health 11: 229–232.

Neukirch, F., C. Ségala, Y. Le Moullec, M. Korobaeff, and M. Aubier. 1998. Short-term effects of low-level winter pollution on respiratory health of asthmatic adults. Archives of Environmental Health 53(5): 320–328.

Pope, C.A., R.T. Burnett, M.J. Thun, E.E. Calle, D. Krewski, K. Ito, and G.D. Thurston. 2002. Lung cancer, cardiopulmonary mortality, and long-term exposure to fine particulate air pollution. Journal of American Medical Association 287(9): 1132–1141.

Pope, C.A., M.J. Thun, M.M. Namboodiri, D.W. Dockery, J.S. Evans, F.E. Speizer, and C.W. Heath, Jr. 1995. Particulate air pollution as a predictor of mortality in a prospective study of U.S. adults. American Journal of Respiratory and Critical Care Medicine 151(3Pt1): 669–674.

Prescott, G.J., G.R. Cohen, R.A. Elton, F.G. Fowkes, and R.M. Agius. 1998. Urban air pollution and cardiopulmonary ill health: a 14.5 year time series study. Occupational and Environmental Medicine 55(10): 697–704.

Roemer, W., G. Hoek, and B. Brunekreef. 1993. Effects of ambient winter air pollution on respiratory health of children with chronic respiratory symptoms. American Review of Respiratory Disease 147(1): 118–124.

Segala, C., B. Fauroux, J. Just, L. Pascual, A. Grimfeld, and F. Neukirch. 1998. Short-term effect of winter air pollution on respiratory health of asthmatic children in Paris. European Respiratory Journal 11(3): 677–685.

Shanghai Municipal Bureau of Public Health. 2000. Shanghai Health Yearbook. Pp. 54–58. Shanghai: Shanghai Science and Technology Literature Publishing House.

Wang, H. 2001. Willingness to Pay for Reducing the Risk of Death by Improving Air Quality: A Contingent Valuation Study in Chongqing, China. Presented at 3rd International Health Economic Association Conference, July 22–25, 2001, York, England.

Wang, J., Q. Wang, and Z. Bi. 1994. A survey on acute respiratory disease. Chinese Journal of Epidemiology 15(3): 141–144.

Wilson, R., and J. Spengler. 1996. Particles in Our Air: Concentrations and Health Effects. Cambridge, Mass.: Harvard University Press.

Wordley, J., S. Walters, and J.G. Ayres. 1997. Short term variations in hospital admissions and mortality and particulate air pollution. Occupational and Environmental Health 54(2): 108–116.

Xu, X., D.W. Dockery, D.C. Christiani, B. Li, and H. Huang. 1995. Association of air pollution with hospital outpatient visits in Beijing. Archives of Environmental Health 50(3): 214–220.

Ye, B., X. Ji, H. Yang, X. Yao, C.K. Chan, S.H. Cadle, T. Chan, and P.A. Mulawa. 2003. Concentration and chemical composition of $PM_{2.5}$ in Shanghai for a 1-year period. Atmospheric Environment 37(4): 499–510.

Zmirou, D., J. Schawartz, M. Saez, A. Zanobetti, B. Wojtyniak, G. Touloumi, C. Spix, A. Ponce de Leon, Y. Le Moullec, L. Bacharova, J. Schouten, A. Ponka, and K. Katsouyanni. 1998. Time-series analysis of air pollution and cause-specific mortality. Epidemiology 9(5): 495–503.